Topics in Low-Dimensional Topology

In Honor of Steve Armentrout

Proceedings of the Conference on Low-Dimensional Topology

Topics in Low-Dimensional Topology

In Honor of Steve Armentrout

Editors

A. Banyaga
H. Movahedi-Lankarani
R. Wells

The Pennsylvania State University

World Scientific
Singapore • New Jersey • London • Hong Kong

Published by

World Scientific Publishing Co. Pte. Ltd.

P O Box 128, Farrer Road, Singapore 912805

USA office: Suite 1B, 1060 Main Street, River Edge, NJ 07661

UK office: 57 Shelton Street, Covent Garden, London WC2H 9HE

British Library Cataloguing-in-Publication Data
A catalogue record for this book is available from the British Library.

TOPICS IN LOW-DIMENSIONAL TOPOLOGY

ISBN 981-02-4050-3

Printed in Singapore by Uto-Print

PREFACE

This volume contains the proceedings of the conference "On Low Dimensional Topology" which was held at the Pennsylvania State University, University Park, in May 1996 in honor of Steve Armentrout.

Recent success with the four-dimensional Poincaré Conjecture has revived interest in low dimensional topology, especially in the three-dimensional Poincaré Conjecture and other aspects of the problem of classifying three-dimensional manifolds. These problems have been a driving force in generating a great body of research and insight.

The main subjects treated in this book include the paper by Valentin Poénaru on the Poincaré Conjecture and its ramifications, giving an insight into the herculean work of the author on the subject. The paper of Steve Armentrout on "Bing's dogbone space" belongs to the topics in three-dimensional topology motivated by the Poincaré Conjecture. Sukhjit Singh gives a nice sythesis of Steve's work and Dennis Sullivan discusses his own deep and far reaching work on metric and conformal gauges.

Also included in this volume are shorter original papers, dealing with somwhat different aspects of geometry, dedicated to Steve by his colleagues: Augustin Banyaga (and Jean-Pierre Ezin), David Hurtubise, Hossein Movahedi-Lankarani, and Robert Wells.

The speakers at the conference were Robert Daverman, Steve Ferry, Cameron Gordon, William Jaco, Valentin Poénaru, Sukhjit Singh, and Dennis Sullivan.

The organizers of the conference wish to thank Penn State Continuing and Distance Education for its financial support, and the Mathematics Department for its logistical and scientific support.

Augustin Banyaga
Hossein Movahedi-Lankarani
Robert Wells
Editors

CONTENTS

INTRODUCTION

During May 1996, a conference in honor of Steve Armentrout was held at Penn State University. As detailed in the Review in this volume, written by his most recent student Sukhjit Singh, the work of Armentrout during the past four decades has been a fundamental propellant for the development of one of the deepest and most formidable disciplines in mathematics, the purely topological theory of manifolds, especially three-dimensional ones. In this review, Singh traces the context, development, and effect of that work showing how the themes, begun with R. L. Moore's E^2-Decomposition Theorem and continued with the Bing Shrinkability Criterion and the Dogbone Decomposition Space, are naturally completed with the Armentrout Cellular Decomposition Theorem and the Straight Arc Decomposition Space

In his review, Singh leads the reader to the Bing Dogbone Decomposition Space and the role, apparently unsatisfactory to Bing, that shrinkability plays in the proof that the quotient is not a manifold. The paper by Armentrout in this collection presents a proof that the quotient is not strongly locally simply connected at certain points. This result implies that the quotient is not a manifold, revealing a more easily understood reason for the quotient to fail to be a manifold. However, Armentrout notes that the proof makes enough use of the techniques of shrinkability so that the new proof itself is not actually easier.

Again in his review, Singh alludes to the underlying but often undeclared motivation for seeking such results as the Armentrout Cellular Decomposition Theorem: The Three-Dimensional Poincaré Conjecture remains open. One worker willing to admit his own investment of energy in the problem is Valentin Poénaru, who contributes a paper to this collection outlining a part of his massive and far-reaching program to settle the question. Although the Poincaré Conjecture is a purely topological and three-dimensional problem, Poénaru deploys his attack in the smooth category among manifolds of dimension four, with the third and final stage in his program being to prove that if for a smooth fake three-cell Δ^3 the space $\Delta^3 \times I$ has no 1-handles, then Δ^3 is the standard three-cell. In this context, Poénaru is able to draw upon a fascinating array of devices, from smoothly wild but topologically tame handles to fundamental groups at infinity.

The paper of Dennis Sullivan also uses more than the purely topological structure of manifolds, but not so much as a smooth structure. Here the seminal concept is that of a Lipschitz structure or metric gauge: a locally bi-Lipschitz equivalence class of metrics. For these, the remarkable theorem holds that any manifold of dimension other than four admits a unique metric gauge locally equivalent to the Euclidean one. Sullivan outlines how these and the similar locally Euclidean conformal gauges admit a theory of forms, Hodge operator and exterior derivative, leading to Atiyah-Singer theory and even Yang-Mills-Donaldson theory in dimension four.

Finally, some papers were contributed to this collection by friends and colleagues of Steve, papers not directly relevant to his interests. One of these, by A. Banyaga and J.-P. Ezin, introduces a new conformal invariant for compact Riemannian manifolds. The authors prove that this invariant is nontrivial if and only if the Riemannian manifold is conformal to the sphere with its standard metric. The paper of D. Hurtubise provides a complete and elementary proof of the "well-known folk theorem" which says that holomorphic maps from $\mathbb{C}P^1$ to $G_{n,n+k}(\mathbb{C})$ are given by equivalence classes of matrices with polynomial entries. This folk theorem is the starting point for several papers by various authors who study the topology of holomorphic mapping spaces. The fact that a holomorphic map from $\mathbb{C}P^1$ to $G_{n,n+k}(\mathbb{C})$ is given locally by a matrix of polynomials is a consequence of Chow's Theorem. The theorem Hurtubise proves (without reference to Chow's Theorem) is an improvement of this local result. The last paper in this collection, by H. Movahedi-Lankarani and R. Wells, extends the classical Hopf-Rinow theorem to a compact non-manifold subset of a C^2 Riemannian manifold by showing that if such a set supports a Borel measure with finite mass-scaling dimension at evry point, then any isometry (with respect to the inherited metric) is C^1, and the group of all such isometries is a Lie group.

Mathematics of Steve Armentrout: A Review

By

S. Singh
Dept. of Math
Southwest Texas State University
San Marcos, Texas 78666

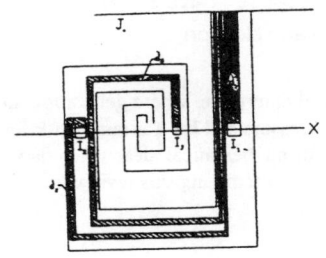

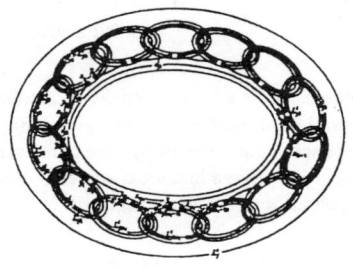

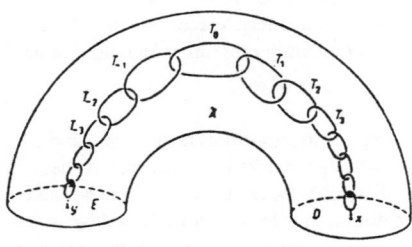

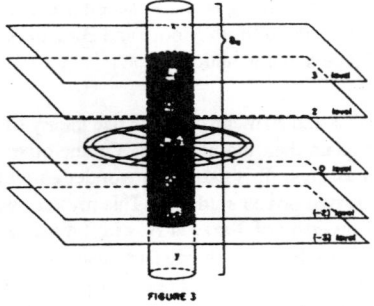

FIGURE 3

1

Mathematics of Steve Armentrout:
A Review

By

S. Singh
Dept. of Math
Southwest Texas State University
San Marcos, Texas 78666

Introduction

"If I have seen farther than others, it is because I have stood
on the shoulders of giants." – Isaac Newton

Mathematics is a human endeavor, a collective ongoing social enterprise, where generation after generation of mathematicians extend the work of their predecessors and lay a groundwork for future scholars. Ostensibly, one gains greater appreciation for mathematical ideas when they are presented within a historical context. With this in mind, we are attempting this review of mathematics of Steve Armentrout.

We surround Steve's mathematics with related framework. We, however, keep this framework to a minimum in order to maintain focus on Steve's works and to gain expediency. The first phase of Steve's mathematics is strongly influenced by, his Ph.D. mentor, the legendary R. L. Moore. The second phase of Steve's mathematics is more directly influenced by R.H. Bing's groundbreaking work on decompositions of the 3-dimensional Euclidean space. We will discuss the works of both Moore and Bing in reasonable detail for coherently presenting and motivating Steve's mathematics.

Bing and Armentrout both use plenty of geometric drawings in their writings to motivate and to explain their mathematics. On the other hand, Moore's published works has a scarcity of these drawings. It is natural to wonder about this difference of styles between the master teacher and his two prized students. This mystery was solved when during my visit to Moore's archive at the University of Texas at Austin, I discovered an abundance of geometric drawings, hand-drawn by Moore himself—he de facto used during the course of his mathematical discoveries. It is not clear why Moore did not incorporate these drawings into his writings. We have generously sprinkled geometric drawings from the works of Steve Armentrout throughout this review. In addition, we have also included several drawings from Bing and Moore whenever appropriate. These drawings add intrigue for the novice and, indeed, they are self-explanatory for an expert.

This review is a mathematical story, a mathematical journey through the mathematics of Moore, Bing, and Armentrout. The current version of this review is based on an earlier presentation given during the topology conference held at the Pennsylvania State University in 1996 to honor Steve's 66-th birthday. Our 1996 presentation was prepared in slide show format using Microsoft Power Point and delivered on a projection screen attached to a PC.

A Few Disclaimers:

- *This is not a comprehensive review but rather a limited perspective of select few ideas manifestly present in Steve's work.*

- *We set up minimal mathematical framework for this review but, nonetheless, we develop peripheral historical framework. We travel forward in time and observe the evolution of mathematical ideas!*

- *This paper is mostly expository and highly non-technical. This review does not catalog numerous developments related to Steve's work. I extend my apologies to colleagues whose noteworthy contributions cannot be acknowledged for the sake of expediency.*

An Overview
of Mathematical Framework

This paper is divided into the following sections for conveniently examining Steve's work within these contexts:

❶ Foundations of Topology *A ´ La* R. L. Moore

❷ USC Decompositions of E^3: "A Golden Age of Examples and Theory (1950-1970)

❸ Local Homotopy Properties

❹ Applications of Decomposition Spaces

❺ Concluding Remarks

Section ❶

Foundations of Topology
A´ La R. L. Moore

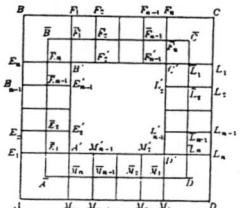

R. L. Moore's (1882-1974) mathematical works have served as a fountainhead for many branches of topology. Moore's research contributions to point-set topology are pinnacles of achievement.

Legends about Moore's method of teaching still echo the classrooms and halls of academia. Moore had 50 Ph.D. students from 1918 to 1969. Moore trained his students by emphasizing didactic rigor, creativity, and discovery within the realm of his favorite discipline, the topology of the plane.

> **R. H. Bing, 18th Ph.D. student of Moore, completed his thesis on "webs" in the plane in 1945.**
>
> **Steve Armentrout, 29th student of Moore, wrote his thesis on "spirals" in the plane to complete his Ph.D. in 1956.**

A Quick Tour Of Moore's Mathematics:

R. L. Moore had an auspicious beginning by publishing his first major paper in 1907:

> *Geometry in which the sum of the angles of every triangle is two right angles*, Trans. Amer. Math. Soc. 8 (1907), 369-378.

This paper was well received as indicated by the following excerpts from its review by Prof. Halstead, one of the leading geometers of the time.

> **"The attention of geometers should be directed to a remarkable article by Dr. R. L. Moore, of Princeton, whose extraordinarily elegant proof of the redundancy of Hilbert's axiom first appeared in the American Mathematical Monthly."**
>
> **"This new article (cited above) is also a perfecting of the work of the Hilbert school, but reaches new results so unexpected, so profound as to be nothing less than epoch making."**
>
> **Reference: Science, Vol. 26, p. 551, October 25, 1907.**

Between 1907 and 1929, Moore published a large number of papers on foundations (fifty to be exact) culminating with his synthetically seminal volume:

> ***Foundations of Point Set Theory***, **AMS Colloquium Publications, Vol. 13, 486 pages, New York, 1932.**
>
>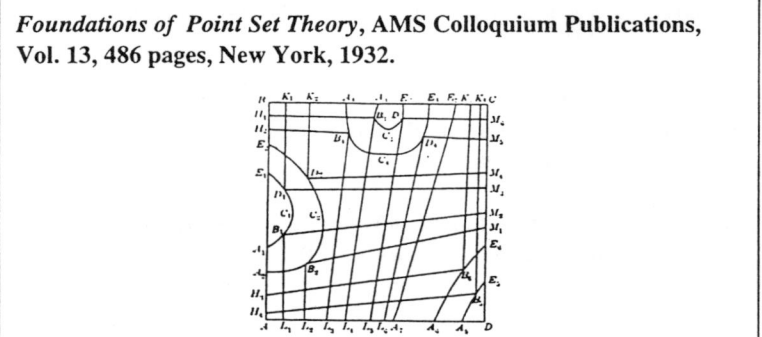

After 1932, Moore expended a considerable amount of energy on pedagogy, 40 of his 50 doctoral students completed their dissertations in this period. He, nevertheless, vigorously studied subsets of the plane while maintaining an ongoing interest in areas heretofore investigated.

After receiving his Ph.D. in 1956 under the direction of R. L. Moore, Steve's research areas were akin to prototypic areas studied by Moore.

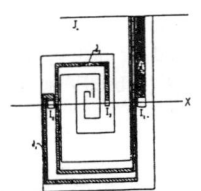

Steve's early publications:

Concerning a certain collection of spirals in the plane, Duke Math. J. 26 (1959), 243-250.

Separation theorems for some plane-like spaces, Trans. Amer. Math. Soc. 97 (1960), 120- 130.

A Moore space on which every real-valued continuous function is constant, Proc. Amer. Math. Soc. 12 (1961), 106-109.

Completing a Moore space, Topology Conference ASU-1967, E. E. Grace, ed., Tempe, Ariz., 22-35.

These papers constitute the first phase of Steve's mathematics. The second phase, based on Moore's concept of "upper-semi-continuous decomposition" and inspired by his theorem, will be explored as this paper progresses.

Definitions and Conventions:

A decomposition G of a space X is a partition of X , i.e., elements of G are pairwise disjoint nonempty subsets of X whose union is X.

For a given decomposition G of a space X, we prefer to denote the quotient set by X/G rather than G and we denote the natural projection from the set X onto the quotient set X/G by: $\pi : X \to X/G$.

Now, convert X/G into a topological space by endowing it with the quotient topology: A subset of U of X/G is declared open if and only if $\pi^{-1}(U)$ is an open subset of X. The quotient space X/G will be called the decomposition space from now on.

Without additional conditions on the decomposition G , the decomposition space X/G is too unwieldy! So, Moore studied the following condition:

A decomposition G of a space X is <u>upper semi-continuous</u> (Abbreviate: <u>usc</u>) if and only if for each g ∈ G and for each open subset U of X with g ⊂ U, there is a saturated open (union of elements of G) subset V with g ⊂ V ⊂ U. Equivalently, G is usc if and only if the projection map is closed.

Remarks and Ongoing Assumptions:

> 1. **All spaces will be at least separable metric.**
>
> 2. **We are only interested in usc decompositions.**
>
> 3. **We are only interested in usc decompositions into continua, i.e., elements of decompositions will be compact and connected.**

Moore's Theorem:

> **If G is an usc decompositions of the plane E^2 into nonseparating continua (E^2 -g is connected for every g in G), then the decomposition space E^2/G is homeomorphic to E^2.**
>
> **Reference:** *Concerning upper semi-continuous collections of continua,* **Trans. Amer. Math. Soc. 27 (1925), 416-428.**

How Does Moore Prove This Theorem?

In his 1916 paper cited below, Moore has given a characterization of E^2 by using his Axioms 1-8. Ostensibly, it was natural for Moore to demonstrate that the decompositions space E^2/G, indeed, satisfied Axioms 1-8. Fait accompli!

Reference: *On the foundations of plane analysis situs*, Trans. Amer. Math. Soc. 17 (1916), 131-164.

Summative Comments:

■ Moore's biggest contribution is crystallization of the concept of "upper semi-continuous decomposition."

■ Ever since its propitious introduction, this concept appears as an essential assumption in numerous theorems!

■ Moore's theorem has provided valuable insight and inspiration for researches ever since its arrival in 1925.

This will become more transparent as we move forward.

Section ❷

USC Decompositions of E^3: A Golden Period of Examples (1950 - 1970)

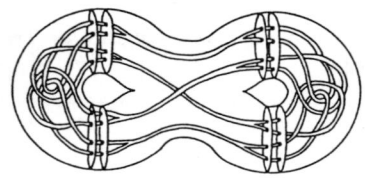

There was an ongoing interest among Moore scholars to extend Moore's work from the plane E^2 to higher dimensional Euclidean spaces. Even Moore, on occasion, forayed into the uncharted territory of higher dimensions; see Moore's paper: [*Concerning upper semi-continuous collections of continua*, Trans. Amer. Math. Soc. 27 (1925), 416-428.]

Ostensibly, E^3 is the first dominion to explore. The following quotes from Bing (1955) puts the study of the plane and the study of E^3 in perspective.

BETWEEN 1925-55: "**Upper semicontinuous decompositions of E^2 were discussed. Examples were given of such decompositions. The plane is a wonderful training ground for graduate students—the problems encountered here develop brain power. However, there has been much study devoted to the plane and simple problems that have not already been attacked are difficult to find in this area. Many of the interesting results about the plane discovered in recent years have dealt with decompositions of the plane and the imbedding of continua in the plane.**"

From – Bing, *What topology is here to stay?*, In "Summary of Lectures and Seminars", Madison (1955), Revised (1958), 25-27, University of Wisconsin. Also, in Bing's Collected Papers.

"On the other hand, E^3 is essentially a virgin forest. For many years the problems were so forbidding that few attacks were successful. Now mathematicians are beginning to venture into the woods. We are developing new tools to cope with the problems encountered there. Simultaneously as we learn about E^3, we learn about collections of objects in E^3, decompositions of E^3. This seems a fertile and fruitful field."

From – Bing, *What topology is here to stay?*, In "Summary of Lectures and Seminars", Madison (1955), Revised (1958), 25-27, University of Wisconsin. Also, in Bing's Collected Papers.

Game Plan: Extend Moore's Theorem to E^3

Recall Moore's Theorem:

If G is an upper semicontinuous decomposition of E^2 such that each element of G is *nonseparating*, then the decomposition space E^2/G is topologically E^2.

This extension appeared very difficult!

First: Between 1925-35, it was not even clear how to state a reasonable analog of this theorem for decompositions of E^3 let alone being able to prove it!

Second: Moore used his characterization of the plane, almost two and half decades of hard-work, to prove his theorem.. It was a tour de force! It was a daunting idea to replicate this program in E^3 !

So what happened?

Whyburn's 1935 Remarks:

Antoine has constructed an arc α in E^3 whose complement, $E^3 - \alpha$, is not homeomorphic to the complement of a point, E^3-{point}.

[J. Math. Pures. 86 (1921), 221-325]

Whyburn observed that if G is an usc decomposition of E^3 whose only nondegenerate element is the Antoine arc α then E^3/G is not homeomorphic to E^3.

[From his 1935 AMS address]

Based on this, Whyburn proposed the following:

Whyburn's 1935 Trial Condition:

Prove or Disprove: If G is an usc decomposition of E^3 such that each element of G is point-like, then $E^3/G \approx E^3$.

This provided a simple and concise potential analog of Moore's Theorem.

We will refer to this trial condition as the Whyburn Question:

$g \in G$ is <u>point-like</u> in E^3 $\Leftrightarrow$ $E^3 - g$ is homeomorphic to E^3 - {point}.

Bing (circa 1955, published in 1957), twenty years after Whyburn raised this question, settled the Whyburn question in the negative. Bing's example, currently known as the Dogbone Space (see figure below), opened the floodgates and released the pent up energy. Here is what Bing proved:

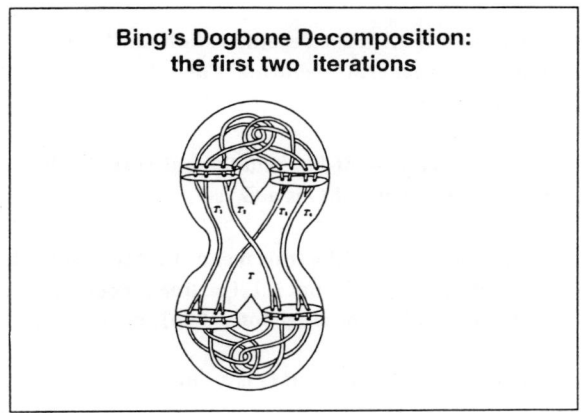

Bing's Dogbone Decomposition:
the first two iterations

Theorem: There exists an usc decomposition G of E^3 such that the elements of G are singletons and tame arcs, and its associated decomposition space E^3/G is not homeomorphic to E^3.

Bing [Ann. of Math. 65 (1957) , 484-500]

In a different paper but in the same volume of Annals, Bing proved the following:

Theorem: Suppose G is an usc decomposition of E^3 with countably many nondegenerate elements. Then, $E^3/G \approx E^3$ provided one of the following holds:

Each element of G is point-like and the union H_G of the nondegenerate elements of G is a G_δ set;

Each element of G is starlike (this means that for each g in G, there is a point p of g such that if L is a line through p, then $L \cap g$ is an interval or the singleton set $\{p\}$; or

Each nondegenerate element of G is a tame arc.

Bing [Ann. of Math. 65 (1957) , 363-374]

The aforementioned 1957 results of Bing on the Dogbone triggered an avalanche of results and ushered in "a golden period of decompositions of E^3" which lasted till 1970(?).

An argument can be made that this golden period de facto began much earlier with Bing's 1952 example [Ann. of Math. 56 (1952), 354-362]. In this paper, Bing proved:

Theorem: Decomposition space S^3/G associated with certain point-like usc decomposition was homeomorphic to S^3.

The study of this 1952 example was a win-win situation for Bing. He would either:

☞ get an example to disprove Whyburn's question,

OR

☞ he would produce an example of a homeomorphism of S^3 of period two whose set of fixed points was the wild 2-sphere described by Alexander.

Whyburn's question waited till the Dogbone Space was conceived by Bing around 1955 by adding more and more complexity to this 1952 example. The following letter from Bing to Moore wistfully announces the discovery of the Dogbone Space:

[Reproduced from the original obtained from Moore Archives at University of Texas, Austin]

March 17, 1955

Dear Professor Moore:

Enclosed is a result that may interest you. I would have preferred that your beautiful theorem for E^2 generalize to E^3 but theorems follow from proofs rather than preferences.

Sincerely,

R. H. Bing

No doubt the 1952 example played a pivotal role on two frontiers. First, theory of group action moved in the direction of smooth topology. Second, it assisted Bing in his discovery of the Dogbone (in 1955).

After the conception of the Dogbone, a large number of mathematicians made numerous contributions within this framework and enriched the theory. We will not be able to mention their results for the sake of expediency.

Nevertheless, we present the following cataclysmic technique as a major byproduct of this era:

Bing's Shrinking Criterion:

Here is a statement of this criterion in a very simple setting:

> Suppose G is an usc decomposition of E^3. The decomposition space E^3/G is homeomorphic to E^3 provided the following "shrinking criterion" is satisfied:
>
> For every open subset U containing the union of the nondegenerate elements and for every $\varepsilon > 0$, there is a homeomorphism of h:$E^3 \to E^3$ such that h restricted to the complement of U is the identity and diam[h(g)] $< \varepsilon$ for every nondegenerate element g .

Intuitively, this is a procedure for gradually shrinking elements to smaller sets without letting other elements grow too large.

A Few Comments on the Shrinking Criterion:

- In its earlier rendition, it appears in Bing's work as early as 1952. After its usefulness was recognized, it became an object of great research activity in its own right.

- Bob Daverman's book [Decompositions of Manifolds, Academic Press Inc., N.Y., 1986] captures many generalizations or improvements and provides reference to previous works.

- The Shrinking Criterion was a homecoming back to Moore's world where the emphasis was on the domain and not on decomposition space.

In another direction, it represented an acute technical paradigm shift: The shrinking criterion circumvents reliance on a characterization of E^3 compared to Moore's *sine qua non* reliance of his characterization of E^2.

But nonetheless, Bing was prepared for the potential use of a characterization of E^3; see [*A Characterization of 3-space by partitionings*, Trans. Amer. Math. Soc. 70 (1951), 15-27].

Contributions of Steve Armentrout in this Setting:

Steve arrived on the scene in the early sixties. A priori, it appears that he was somewhat late! From a historical retrospective, he appeared to have missed the initial excitement surrounding the Dogbone and many other discoveries of the decade of the Fifties.

Steve publishes his first paper on decompositions of E^3 in 1963. Here is the reference:

> *Upper Semi-continuous Decompositions of E^3 with at Most Countably Many Nondegenerate Elements*, Ann. of Math. 78 (1963), 605-618.

This technical paper is a harbinger of things to come. As follow up, Steve publishes the following additional papers:

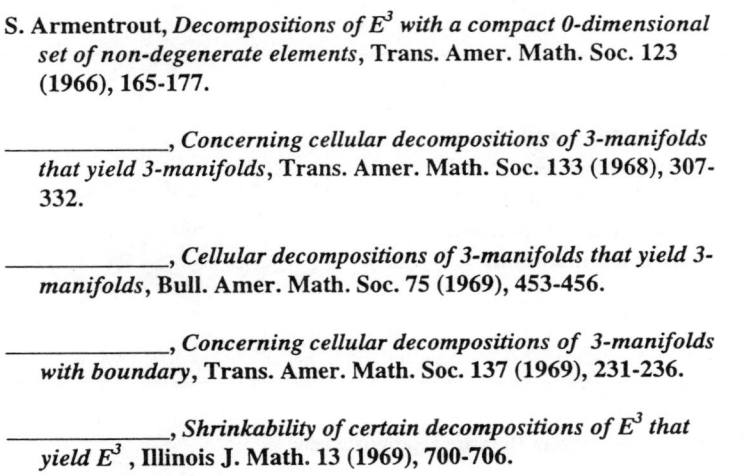

> S. Armentrout, *Decompositions of E^3 with a compact 0-dimensional set of non-degenerate elements*, Trans. Amer. Math. Soc. 123 (1966), 165-177.
>
> _____, *Concerning cellular decompositions of 3-manifolds that yield 3-manifolds*, Trans. Amer. Math. Soc. 133 (1968), 307-332.
>
> _____, *Cellular decompositions of 3-manifolds that yield 3-manifolds*, Bull. Amer. Math. Soc. 75 (1969), 453-456.
>
> _____, *Concerning cellular decompositions of 3-manifolds with boundary*, Trans. Amer. Math. Soc. 137 (1969), 231-236.
>
> _____, *Shrinkability of certain decompositions of E^3 that yield E^3*, Illinois J. Math. 13 (1969), 700-706.

These papers culminate in the following "Capstone Theorem."

Steve Armentrout's Capstone Theorem:

Suppose M is a 3-manifold and G is a
cellular decomposition of M such
that the decomposition space M/G is
a 3-manifold. Then, M/G is
homeomorphic to M.

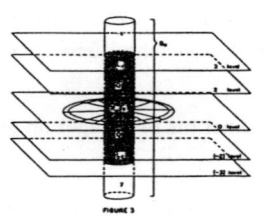

Reference: *Cellular decompositions of 3-manifolds that yield 3-manifolds*,
Memoir AMS 107 (1971), 1-72.

This theorem has numerous corollaries. We will be content to mention the following in
conjunction with Bing's Shrinking Criterion:

Corollary: A cellular usc
decomposition G of a 3-
manifold is shrinkable if and
only if the decomposition
space is a 3-manifold.

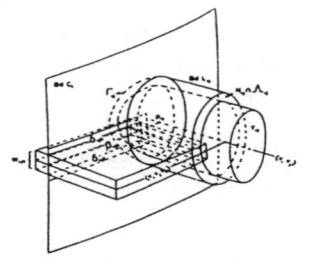

The following comments from Bing are appropriate here:

"Theorems that are stated in a simple and concise fashion are easier to remember and are more likely to survive than statements given in a complicated and involved fashion."

From – Bing [*What topology is here to stay?* 1955]

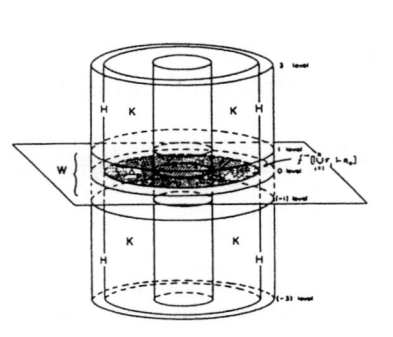

Ostensibly, Armentrout's Capstone Theorem provides a memorable extension of Moore's theorem.

What next for encore?

Point-like Decompositions Revisited:

For manifolds the concept of *cellular decomposition* is the correct analog of the concept of *point-like decomposition* for E^n. Indeed, for E^n these two concepts are identical.

Bing [Point-like decompositions of E^3, Fund. Math. 50 (1962), 431-432] summarizes earlier developments in point-like decompositions and gives new examples. Most importantly, he proposed a decomposition of E^3 into points and segments or straight-arcs in conjunction with the following question that he had raised earlier in 1955:

Bing's 1955 Question:

Does there exist an usc decomposition of E^3 into points and straight arcs (segments) with the decomposition space topologically different from E^3?

An affirmative answer to this problem will provide a cutting edge example in the theory of point-like decomposition of E^3. This leads us to:

Steve Armentrout's Quintessential Example:

Steve publishes his example in 1970, roughly 15 years after the question is posed. Here is the reference:

> **A Decomposition of E^3 into Straight Arcs and Singletons,**
> **Dissertationes Mathematicae, Warszawa, 1970, 1-46.**
>
> **Another related paper is:**
> **A Three -Dimensional Spheroidal Space That Is Not a**
> **Sphere, Fund. Math. 68 (1970), 183-186.**

We will next briefly review several key properties of this decomposition.

- **There exist two parallel planes P and Q such that each nondegenerate element of the decomposition has one end on P and the other end on Q.**

- **The decomposition has a Cantor set of nondegenerate elements.**

- **Each point in the associated decomposition space has arbitrarily small neighborhoods which are bounded by 2-spheres.**

- **Moreover, each point has arbitrarily small neighborhoods which are compact absolute retracts with 2-sphere boundaries.**

Its higher dimensional analog is not known yet! What next?

Section ❸

Local Homotopy Properties

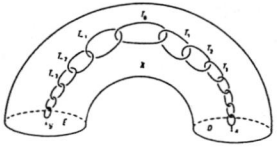

Steve studied several homotopy theoretic properties called "UV" properties of decomposition spaces. Here are a few of his papers in this area:

Decompositions into compact sets with UV properties, (joint work with Thomas M. Price), Trans. Amer. Math. Soc. 141 (1969), 433-442.

UV properties of compact sets, Trans. Amer. Math. Soc. 143 (1969), 487-498.

Homotopy properties of decompositions, Trans. Amer. Math. Soc. 143 (1969), 499-507.

These papers are very general in nature. They capture and crystallize many theoretical ideas that arose within the framework of decomposition spaces. There was a considerable interest in these ideas vis-à-vis theory of retracts, generalized manifolds, and later in shape theory.

We will not fully develop these ideas in this paper for the sake of expediency but, nevertheless, we discuss the following somewhat related local homotopy theoretic property by first discussing toroidal decompositions:

Toroidal Decompositions: Why Were They Studied?

Because an ongoing agenda
item was:

Simplify examples and give
understandable proofs!

[Bing, 1955]

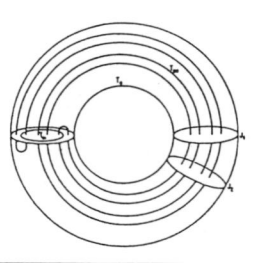

Ever since the auspicious arrival of the Dogbone, there was a keen interest to simplify Bing's proof as well as seek alternative examples towards the same end. Bing, himself, led the charge. We quote from Bing's article [*Decompositions of E^3*, Toplogy of 3-Manifolds and Related Topics (M. K. Fort, Jr., editor), 5-21, Prentice-Hall; Englewood Cliffs, N.J. 1962]:

"To prove that two spaces are topologically different, it is
convenient to discover some simple property that one of them has
but the other does not. It is less elegant to prove the lack of a
homeomorphism. This less elegant method was used to show that the
Dogbone space was topologically different from E^3."

"I had hoped that others writing about the Dogbone Space would
give a more elegant proof that it was topologically different from E^3.
Several have not availed themselves of this opportunity."

Steve's Response to Bing:

Definition: A space X is *strongly locally simply connected* if and only
if every point of X has arbitrarily small simply connected open
neighborhoods.

Steve's paper on toroidal decompositions cited below proves:

Theorem: There exists a point-like toroidal decomposition of E^3 such that the decomposition space E^3/G is not strongly locally simply connected, and thus, different from E^3. Moreover, this method produces infinitely many topologically distinct similar examples.

On the strong local simple connectivity of the decompositions spaces of toroidal decompositions, Fund. Math. **69** (1970), 15-37.

We point out, however, that Steve had earlier written the following joint paper with Bing, but the method of proof was different:

A toroidal decomposition of E^3, Fund. Math. **60** (1967), 81-87.

Moreover, Bing [Fund. Math 50 (1962), 431-453] had also defined a torodial decomposition with the same goals, but again the proof did not use strongly local simple connectedness.

Aside from Whyburn's age old question, other applications of toroidal decompositions became prevalent. Others contributed to the theory as well but we will not be able to sort this out in this paper.

There still was a loose end!

Recall Bing's plea for a simpler proof establishing that the Dogbone Space is not homeomorphic to E^3. Let us note:

Steve has recently submitted his long awaited proof that the Dogbone Space, Bing's original rendition, is not strongly locally simply connected.

This is the type of proof Bing was asking for.

So, a promise is fulfilled! But wait, there is more to go!

Steve Armentrout's Knotted Dogbone Spaces:

In Steve's paper cited below, the following appears:

> - **Knotted Dogbone Spaces are defined.**
>
> - **The main result of this paper is that Knotted Dogbone Spaces are not strongly locally simply connected.**

By combing this result with previous results, we have the following synthesis of several decades of work:

> **The strong local simple connectivity is used to distinguish between E^3 and the point-like decomposition spaces associated with:**
>
> - **classical Dogbone decomposition,**
> - **knotted dogbone decompositions, and**
> - **toroidal decompositions**
>
> *Local properties of knotted Dogbone Spaces*, **Top. And Its Applications 24 (1986), 41-42.**

What next?

Section ❹

Applications of Decompositions

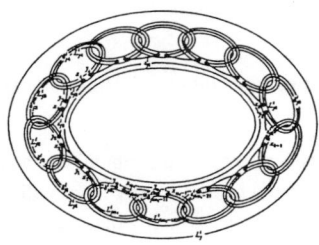

As a confluence of Bing's savvy with decompositions and Borsuk's skills with retracts, the following joint paper emerged:

> **Bing & Borsuk,** *A 3-dimensional absolute retract which does not contain any disk,* **Fund. Math. 54 (1964), 159-175**

More specifically, they proved that there was an usc decomposition of the closed 3-ball into a null collection of arcs whose associated decomposition space did not contain any 2-cell (disk).

As a consequence of their techniques, it followed among other things, that there are generalized 3-manifolds having the homotopy type of S^3 which do not contain 2-cells. Moreover,

---

Bing and Borsuk asked:

Whether their decomposition space did not contain any 2-dimensional compact absolute retract as well?

A formidable problem!

Steve provided an affirmative answer in the following monograph:

> *A Bing-Borsuk retract which contains a 2-dimensional absolute retract*, **Dissertationes Math., Warszawa, 1975, 1-39.**

Another highly technical and difficult monograph!

In conclusion, we mention Steve's passion with the 3-Dimensional Poincare' Conjecture.

Over the years, Steve has vigorously worked on this ubiquitous conjecture. Spin off from this work is beginning to appear in print and he is currently more willing to publish numerous compendium of ancillary results. As per our ongoing disclaimer, we will not be able to review this work (still in progress at this writing).

Section ❺

Concluding Remarks

We quote from the following excerpts from Bing's letter, dated April 28, 1971 to R. L. Moore [Letter obtained from Moore's Archives at the University of Texas, Austin]:

> **"Armentrout writes very long papers. My preprint file of his papers bulges..."**
>
> **"Armentrout is an excellent teacher. He spent one year at Wisconsin and was impressed with the enthusiasm he engendered among the students who liked to tackle hard problems. He has an excellent publication record. I think very highly of his work and influence."**

Back to the Future:

We now quote excerpts from R. L. Moore's letter of recommendation, dated Dec. 13, 1955, to Professor Knowler (Univ. of Iowa) [copy obtained from Moore's archives at the University of Texas, Austin]:

> **"During our phone conversation earlier this year, I spoke of Mr. Steve Armentrout. He was born in Eldorado, Texas, June 19, 1930 and entered the University of Texas in September, 1947..."**
>
> **"At our Commencement in June, 1951, 371 individuals received the degree of Bachelor of Arts. Of these 371, ... only 4 received it with highest honors. Mr. Armentrout was one of these 4."**
>
> **"I think his dissertation will be concerned with a solution which he has obtained of what I considered to be a decidedly difficult problem concerning spirals. He has also been attacking other problems and I certainly think he will continue to produce. He is a man of high ideals, with a personality that is in his favor...."**

Steve Armentrout as a Teacher:

Like the great master teacher R. L. Moore, Steve has continually carried sound didactic principles, enthusiasm, and a sense of discovery to every classroom – undergraduate and graduate alike. If my count is correct, Steve had 12 Ph.D. students. Here is a list:

Ph.D. Students of Steve Armentrout:

Joseph M. Martin **(1962 at Iowa)** **Univ. of Wyoming**	**Bruce A. Anderson** **(1966 at Iowa)** **Arizona State University**
Donald V. Meyer **(1962 at Iowa)** **Central College**	**Thomas C. Hutchinson** **(1966 at Iowa)** **University of Cincinnati**
Orabi H. Alzoobaee **(1962 at Iowa)** **Univ. of Baghdad**	**William L. Voxman** **(1968 at Iowa)** **University of Idaho**
Lloyd L. Lininger **(1964 at Iowa)** **SUNY at Albany**	**John P. Neuzil** **(1969 at Iowa)** **Kent State University**
J. Brauch Fugate **(1964 at Iowa)** **University of Kentucky**	**R. Richard Summerhill** **(1969 at Iowa)** **Kansas State University**
Richard M. Schori **(1964 at Iowa)** **Oregon State University**	**Suji Singh** **(1973 at Penn State)** **SW Texas State Univ.**

Service to the Mathematical Community:

Between R. L. Moore and Steve Armentrout, there is almost a century of continued contributions to mathematical research and ongoing service to the scientific community at large.

- **Steve served as an Editor of Transactions of the American Mathematical Society, 1970-73.**

- **From 1970-1993, he served on the AMS Council, AMS Associate Treasure, and member of the AMS Board of Trustees.**

Steve discharged all of these responsibilities happily and judiciously. Steve has always been a great team player. He served "the decomposition team" by writing timely survey articles and by performing numerous other services.

It is difficult to do a comprehensive review of Steve's mathematics due to multitudes of themes and ideas compounded with his current research endeavors– I have shared only a select few thematic threads.

Steve's papers and his presentations are often filled with fascinating drawings to motivate, to inspire, as well as to clarify complex conceptual frameworks. Many of his drawings are already sprinkled throughout this paper to create an ambiance of Steve's persona as well as to intrigue the reader not familiar with Steve's work. As a fitting tribute, we conclude this presentation with a graphical display from Steve's straight-arc decomposition.

Section ⑥

A Graphical Display From Steve's "Straight-Arcs" Example

A decomposition of E^3 into straight arcs and singletons

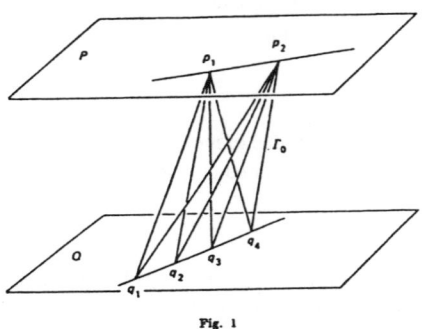

Fig. 1

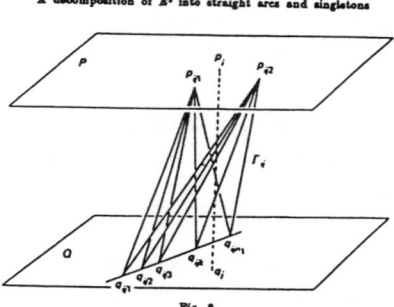

Fig. 2

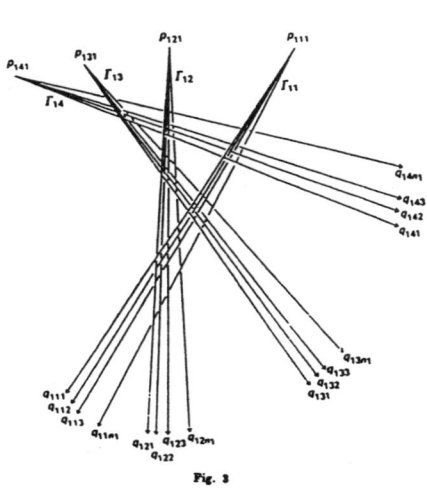

Fig. 3

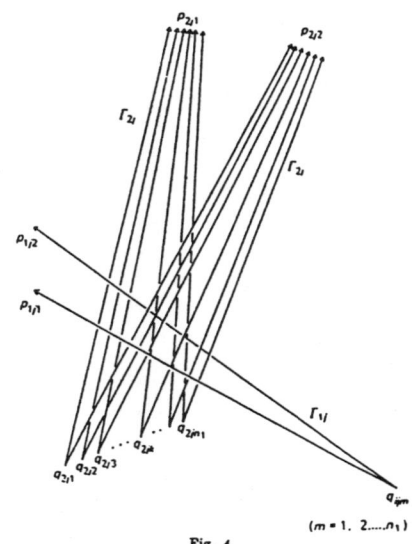

$(m = 1, 2....n_1)$

Fig. 4

3. Description of G

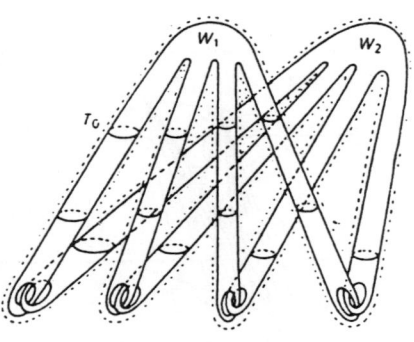

Fig. 5

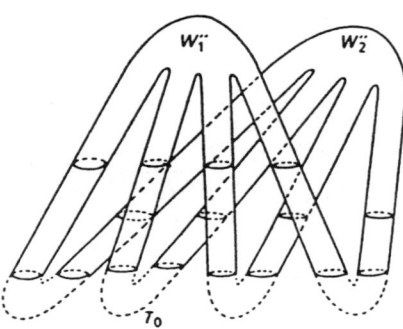

Fig. 6

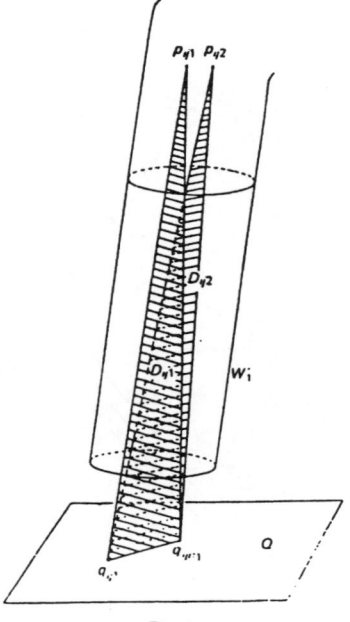

Fig. 7

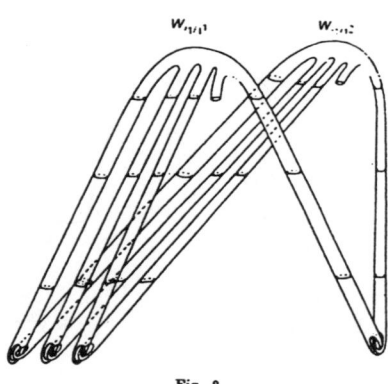

Fig. 8

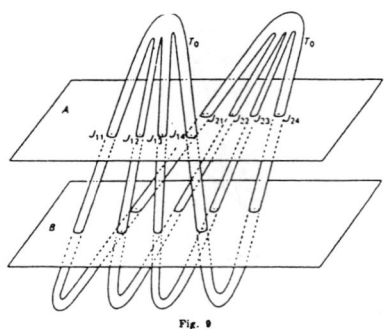

Fig. 9

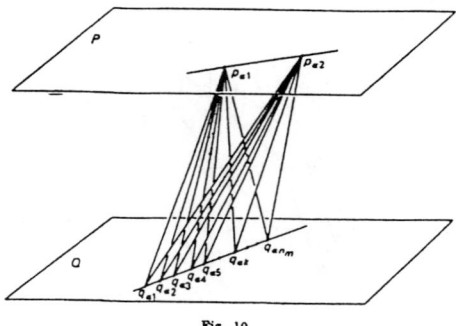

Fig. 10

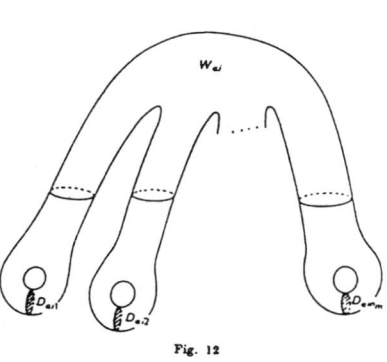

Fig. 12

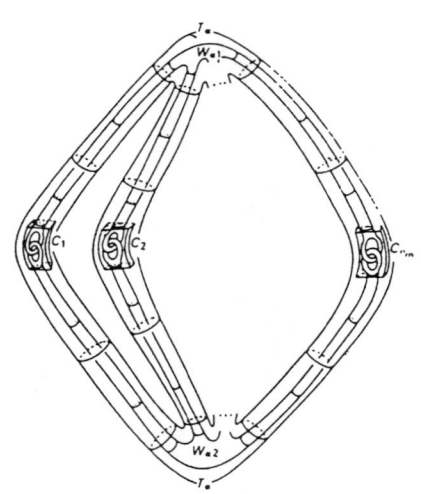

Fig. 17

BING'S DOGBONE SPACE IS NOT STRONGLY
LOCALLY SIMPLY CONNECTED

STEVE ARMENTROUT

1. Introduction

The main result of this paper is that there are points of Bing's dogbone decomposition space which do not have arbitrarily small simply connected open neighborhoods. This answers a question raised by Bing in 1955 [7].

Bing's dogbone decomposition, described in [8], was the first example of a cellular decomposition of euclidean 3-dimensional space $\mathbb{R}^3$ whose decomposition space is topologically distinct from $\mathbb{R}^3$. Bing established this result by showing that the dogbone decomposition is not shrinkable.

Bing asked ([7], [8], [10]) for a local topological property that distinguishes the dogbone space from $\mathbb{R}^3$. The dogbone space is an absolute neighborhood retract, and hence is locally simply connected at each point. Bing [10] raised the question as to whether each point of the dogbone space has arbitrarily small simply connected open neighborhoods; see also [11]. If X is a topological space, then X is *strongly locally simply connected* if and only if each point of X has arbitrarily small simply connected open neighborhoods.

In this paper, we shall prove that Bing's dogbone space is not strongly locally simply connected. There have been some related results concerning spaces obtained by varying the dogbone construction ([5], [13]). Strong local simple connectivity of certain toroidal decomposition spaces was studied in [1].

In this paper, as in others ([1], [5], [13], [14]) on local properties of decomposition spaces, the arguments involve the ideas used to establish nonshrinkability, and hence do not provide alternative proofs of topological distinctness of original and decomposition space. In addition, the proofs about local properties are often significantly more complicated. For these reasons, the study of local properties of decomposition spaces has not yet provided methods of first choice for distinguishing topologically between the original and the decomposition space.

However, the study of local properties of decomposition spaces is of value. This has, for example, been shown in the uses of cell-like decompositions to study retracts ([4]). A second application is to questions concerning extensions of decompositions. In [6], we give an application of this type. A third application is to the problem of topological characterization of 3-manifolds. Examples such as [2] show that a space may have much of the characteristic local structure of a 3-manifold, and yet not be one [3].

In Section 2, we give a brief description of the dogbone decomposition. The main technical result of this paper is a generalization of Theorem 11 of [8]. This result is established in Section 12, and Sections 3–12 give preliminary lemmas for this result. The proof of the main result is given in Section 13.

2. DESCRIPTION OF THE DECOMPOSITION

Let T_0 be a polyhedral solid double torus in $\mathbb{R}^3$ as shown in Figure 1. Let T_1, T_2, T_3, and T_4 be four mutually disjoint polyhedral solid double tori in Int T_0 obtained by thickening slightly the four polyhedral graphs shown in Figure 1. For any two of these graphs, the upper loops are linked, as are the lower ones. There are disjoint rectangular 3-cells B_0 and C_0 in Int T_0, as indicated in Figure 1, B_0 containing the upper, and C_0 the lower, loops of the graphs.

The graphs indicated above are the *cores* of the four T's. For each i, we let L_i denote the core of T_i. For each i, L_i lies in a vertical plane in $\mathbb{R}^3$, and no three of these planes intersect. We assume that for each i, the upper and lower loops of L_i are rectangles with horizontal and vertical sides. The upper loops are situated in B_0 as shown in Figure 2. A similar result holds for C_0. For each i, the stem of L_i is an arc which intersects each horizontal plane in at most one point.

If $i = 1, 2, 3$, or 4, then in T_i, we construct four mutually disjoint polyhedral solid tori T_{i1}, T_{i2}, T_{i3}, and T_{i4}, related to T_i as T_1, T_2, T_3, and T_4 are related to T_0. The T_{ij} have polyhedral cores L_{ij} with properties analogous to the cores of the T_i. In T_i, we construct disjoint rectangular 3-cells B_i and C_i related to T_i and the T_{ij} as B_0 and C_0 are to T_0 and the T_i, and in addition, $B_i \subset$ Int B_0 and $C_i \subset$ Int C_0.

In an analogous manner, we define solid double tori T_{ijk} and 3-cells B_{ij} and C_{ij}.

By an *index* we shall mean either 0 or a finite string of symbols, each either 1, 2, 3, or 4. 0 is the only *stage* 0 index, and if $n > 0$, the *stage n* indexes are those of exactly n symbols. If $n > 0$, α is a stage n index, and $i = 1, 2, 3$, or 4, then αi is the stage $(n + 1)$ index obtained by adjoining i to the symbols of α. If $\alpha = 0$, then $\alpha i = i$.

Suppose α is an index and T_α is defined. We construct four mutually disjoint polyhedral solid tori $T_{\alpha 1}$, $T_{\alpha 2}$, $T_{\alpha 3}$, and $T_{\alpha 4}$ in Int T_α, related to T_α as T_1, T_2, T_3, and T_4 and related to T_α as T_1, T_2, T_3, and T_4 and related to T_0, and having polyhedral cores with properties analogous to the cores of the T_i. We also construct rectangular 3-cells B_α and C_α in Int T_α, related to T_α and the four $T_{\alpha i}$ as B_0 and C_0 are related to T_0 and the four T_i. We also require that if $i = 1, 2, 3$, or 4, $B_{\alpha i} \subset$ Int B_α and $C_{\alpha i} \subset$ Int C_α.

For each positive integer n, we may define solid double tori, cores, and 3-cells of the nth stage satisfying conditions indicated above. Let $A = \bigcap\limits_{n=1}^{\infty} (\cup\{T_\alpha : \alpha$ is a stage n index $\})$. Then A is a compact set in $\mathbb{R}^3$. Let G be the decomposition of $\mathbb{R}^3$ consisting of the components of A and singleton subsets of $\mathbb{R}^3 - A$. Since A is compact, then G is upper semicontinuous. It is easily proved that each element of G is cellular in $\mathbb{R}^3$.

G is *Bing's dogbone decomposition* of $\mathbb{R}^3$, and the associated decomposition space $\mathbb{R}^3/G$ is *Bing's dogbone space*. Note that each nondegenerate element of G is an arc which intersects each horizontal plane of $\mathbb{R}^3$ in at most one point.

Our description is based on [8] and [12].

3. THE GENERALIZED BING LEMMA

In the main step, Theorem 11 of [8], of his proof that the dogbone decomposition is nonshrinkable, Bing proved the following result about the discs D and E of Figure 1. *If $h : T_0 \to T_0$ is any homeomorphism with $h|BdT_0$ the identity map, then some element g of G intersects both $h^{-1}(D)$ and $h^{-1}(E)$.*

Bing's argument for this result proves somewhat more. We shall say that a polygonal simple closed curve J in $\mathbb{R}^3$ disjoint from T_0 it links the upper (lower) handle of T_0 if J links the upper (lower) loop of the core of T_0. In this paper, we shall use linking with integer coefficients. Bing's argument for Theorem 11 of [8] establishes the following.

Bing Lemma. *If α is any index, D and E are any polyhedral discs in $\mathbb{R}^3$ such that $Bd\,D$ links the upper handle of T_α, $Bd\,E$ links the lower hand of T_α, and $D \cap E \cap T_\alpha = \emptyset$, then some element of G in T_α intersects both D and E.*

The major part of this paper is devoted to a proof of a generalization of the Bing Lemma in which the discs D and E are replaced by singular discs.

Let Δ_0^2 be a fixed triangular disc in the euclidean plane $\mathbb{R}^2$. Suppose W^3 is a triangulated 3-manifold. Then Δ is a *polyhedral singular disc* in W^3 if and only if Δ is the image of Δ_0^2 under a piecewise linear map $f = \Delta_0^2 \to W^3$ such that (1) $f|Bd\Delta_0^2$ is an embedding and (2) except for a finite number of points, f is locally a homeomorphism. Any such piecewise linear map $f : \Delta_0^2 \to W^3$ is a *defining map for* Δ. We define the *boundary of Δ*, $Bd\,\Delta$, to be $f(Bd\,\Delta_0^2)$; it is a polygonal simple closed curve. If p is a point of Δ_0^2 at which $f : \Delta_0^2 \to W^3$ fails to be a local homeomorphism, then $f(p)$ is an *exceptional point* of Δ. We assume that in all cases, each exceptional point of Δ is a vertex of Δ (as a complex).

Suppose Δ is a polyhedral singular disc in $\mathbb{R}^3$ with a defining map $f : \Delta_0^2 \to \Delta$. Suppose M^2 is a polyhedral 2-manifold with or without boundary, in $\mathbb{R}^3$, $Bd\,\Delta$ is disjoint from M^2, and Δ and M^2 are in relative general position. Then each component of $f^{-1}(M^2)$ is either (1) an arc in Δ_0^2 with both endpoints on $Bd\,\Delta_0^2$ and otherwise lying in Int Δ_0^2, or (2) a simple closed curve lying in Int Δ_0^2.

The statement that α is an *arc of intersection* of Δ with M^2 means that for some arc component α_0 of $f^{-1}(M^2)$, $\alpha = f(\alpha_0)$. Arcs of intersection are singular arcs in M^2 with their end points on $Bd\,M^2$ and which otherwise lie in Int M^2.

The statement that γ is a *curve of intersection* of Δ with M^2 means that for some simple closed curve component γ_0 of $F^{-1}(M^2)$, $\gamma = f(\gamma_0)$. Curves of intersection are loops lying in Int M^2.

Suppose γ is a curve of intersection of Δ with M^2, and γ_0 is the component of $f^{-1}(M^2)$ such that $\gamma = f(\gamma_0)$. Then $\gamma \sim 0$ *on* M^2, or γ *is* (homotopically) *trivial* on M^2, if and only if $f|\gamma_0 : \gamma_0 \to M^2$ is homotopic to 0 in M^2.

Generalized Bing Lemma. *Suppose α is an index, and D and E are any polyhedral singular discs in $\mathbb{R}^3$ such that $Bd\,D$ links the upper handle of T_α, $Bd\,E$ links the lower handle of T_α, and D and E do not intersect in T_α. Then some element of G in T_α intersects both D and E.*

4. Admissible Modifications

In our proof of the Generalized Bing Lemma, we shall need to be able to simplify the ways singular discs intersect the solid double tori and 3-cells of the construction of G. There are, however, restrictions on the types of simplifications that we can make.

Suppose α is an index and D is a polyhedral singular disc in $\mathbb{R}^3$ such that $(Bd\,D) \cap T_\alpha = \emptyset$, and D and $Bd\,T_\alpha$ are in relative general position. A polyhedral singular disc D' in $\mathbb{R}^3$ is *obtained from D by a replacement in T_α* if and only if there is a subdisc Δ of D such that $Bd\,\Delta \subset$ Int T_α and $D' = (D - \Delta) \cup \Delta'$ where Δ' is a polyhedral singular disc in Int T_α, with boundary $(Bd\,\Delta)$. The pair (Δ, Δ') is the *replacement pair*. Note that $D' - \Delta' \subset D$ and hence $D' - \text{Int } T_\alpha \subset D$.

The statement that D' is on *admissible modification of D relative to T_α* means that D' is a polyhedral singular disc in $\mathbb{R}^3$ and there is a finite sequence

$$D = D_0,\ D_1,\ D_2,\ \cdots,\ D_k = D'$$

such that if $i = 0, 1, 2, \cdots$, or $k-1$, then D_i is a polyhedral singular disc in $\mathbb{R}^3$ such that D_{i+1} is obtained from D_i by a replacement in T_α.

Lemma 4.1. *Suppose α is an index, D and E are polyhedral singular discs in $\mathbb{R}^3$ such that $Bd\,D$ and $Bd\,E$ are disjoint from T_α, and each of D and E is in general position relative to $Bd\,T_\alpha$. Suppose D' and E' are polyhedral singular discs in $\mathbb{R}^3$ which are admissible modifications of D and E, respectively, relative to T_α, and D'' and E'' are polyhedral singular discs in $\mathbb{R}^3$ such that D'' and E'' are admissible modifications of D' and E', respectively, relative to T_α. Then D'' and E'' are admissible modifications of D and E, respectively, relative to T_α.*

Lemma 4.2. *Suppose α is an index, D is a polyhedral singualr disc in $\mathbb{R}^3$ such that $(Bd\,D) \cap T_\alpha = \emptyset$, and D and $Bd\,T_\alpha$ are in relative general position. Suppose that if $i = 1, 2, 3$, or 4, D_i is an admissible modification of D relative to $T_{\alpha i}$. Then there exists an admissible modification D' of D such that (1) $(D' - \cup_{i=1}^{4} T_{\alpha i}) \subset D$ and (2) if $i = 1, 2, 3$ or 4, $D' \cap T_{\alpha i} \subset D_i$.*

Proof. If $n = 1, 2, 3$, or 4, there is a finite sequence

$$D = D_{n0}, D_{n1}, \cdots, D_{np_n} = D_n$$

such that if $t = 1, 2, \cdots$, or p_n, then D_{nt} is obtained from $D_{n(t-1)}$ by a replacement in $T_{\alpha n}$. Let $(\Delta_{nt}, \Delta'_{nt})$ be the replacement pair.

Suppose $f : \Delta_0^2 \to D$ is a defining map for D, and $n = 1, 2, 3$, or 4.

Let $F_{n1} = f$, and let δ_{n1} be the subdisc of Δ_0^2 such that $f_{n0}(\delta_{n1}) = \Delta_{n1}$. If $t = 1, 2, \cdots$, or p_n, and f_{nt} is defined, let δ_{nt} be the subdisc of Δ_0^2 such that $f_{n(t-1)}(\delta_{nt}) = \Delta_{nt}$. Then $f_{np_n} : \Delta_0^2 \to D_n$ will be a defining map for D_n. Note that if $1 \leq t \leq p_n$, then $f_{nt}|(\Delta_0^2 - \cup_{s=1}^{t} \delta_{ns}) = f|(\Delta_0^2 - \cup_{s=1}^{t} \delta_{ns})$.

If $n = 1, 2, 3$, or 4, and $1 \leq t \leq p_n$, let $X_{nt} = \delta_{n1} \cup \delta - n2 \cup \cdots \cup \delta_{nt}$. Clearly $X_{n1} \subset X_{n2} \subset \cdots \subset X_{np}$. By induction, if $1 \leq t \leq p_n$, then

$$f_{nt}\big|\Delta_0^2 - X_{nt} = f\big|\Delta_0^2 - X_{nt}\ .$$

To construct D', we shall begin with D_1, modify it in $T_{\alpha 2}$ to obtain D'_2, then modify D'_2 in $T_{\alpha 3}$ to obtain D'_3, and finally modify D'_3 in $T_{\alpha 4}$ to obtain D'.

Consider D_1 and the defining map $g_{20} = f_{1p_1}$ for D_1. Let $D'_{20} = f_{1p_1}(\Delta_0^2) = D_1$. We first shall consider the location of δ_{21} relative to X_{1p_1}. We will show that if X_{1p_1} intersects $Bd\,\delta_{21}$, then $\delta_{21} \subset X_{1p_1}$. Let s be the smallest t such that δ_{1t} intersects $Bd\,\delta_{21}$. We will show that $\delta_{21} \subset \delta_{1s}$.

Suppose not. Note that $\delta_{21} \cap X_{1(s-1)} = \emptyset$ and since δ_{21} is not a subset of δ_{1s}, δ_{21} is not a subset of X_{1s}. Since δ_{1s} intersects $Bd\,\delta_{21}$, then there is a point x of $Bd\,\delta_{21}$ not in X_{15}, and there is a point y of $Bd\,\delta_{21}$ in X_{15}. Since $f_{1s}(X_{1s}) \subset T_{\alpha 1}$ then $f_{1s}(y) \in T_{\alpha 1}$. Since $x \in Bd\,\delta_{21}$ and $x \notin X_{1s}$, then $f_{1s}(x) = f(x)$, and

$f(Bd\,\delta_{21}) \subset T_{\alpha 2}$. But since $Bd\,\delta_{21}$ is connected, and $T_{\alpha 1}$ and $T_{\alpha 2}$ are disjoint, this is a contradiction. Thus $\delta_{21} \subset X_{1s}$, and hence $\delta_{21} \subset X_{1p_1}$.

We may now modify g_{20} to obtain g_{21}. Suppose D_{21} is obtained from D by a modification in $T_{\alpha 2}$, and consider δ_{21}. If X_{1p_1} intersects $Bd\,\delta_{21}$, then $\delta_{21} \subset X_{1p_1}$, and we let $g_{21} = g_{20}$ and define $Y_{21} = \emptyset$. If X_{1p_1} and $Bd\,\delta_{21}$ are disjoint, define g_{21} so that $g_{21}|\delta_0^2 - \delta_{21} = g_{20}|\Delta_0^2 - \delta_{21}$, and $g_{21}|\delta_{21} = f_{20}|\delta_{21}$. This map is well-defined, since in this case, $g_{21}|Bd\,\delta_{20} = f|Bd\,\delta_{20} = f_{21}|Bd\,\delta_{20}$. In this case, let $Y_{21} = \delta_{21}$. Let $D'_{21} = g_{21}(\Delta_0^2)$.

We also need to define X_{21}. From this point on, no sets are added to X_{1p_1} but some may be deleted. We may show that if $1 \leqq t \leqq p_1$ and δ_{21} intersects $Bd\,\delta_{1t}$, then $\delta_{1t} \subset \delta_{21}$. In that case, we delete δ_{1t} from X_{1p_1}. Then X_{21} is obtained from X_{1p_1} by making all such deletions.

It is clear that X_{21} and Y_{21} are disjoint, since $g_{21}(X_{21}) \subset T_{\alpha 1}$ and $g_{21}(Y_{21}) \subset T_{\alpha 2}$.

In a similar manner, we modify g_{21} to obtain g_{22}. If D_{22} is obtained by a modification in $T_{\alpha 2}$, we first consider the location of δ_{22} relative to X_{21}. If $\delta_{21} \subset X_{21}$, then $g_{22} = g_{21}$ and $Y_{22} = Y_{21}$. If X_{21} and $Bd\,\delta_{22}$ are disjoint, define g_{22} so that $g_{22}|\Delta_0^2 - \delta_{22} = g_{21}|\Delta_0^2 - \delta_{22}$ and $g_{22}|\delta_{22} = f_{21}|\delta_{22}$. In this case, $Y_{22} = Y_{21} \cup \delta_{22}$. Let $D'_{22} = g_{22}(\Delta_0^2)$. We define X_{22} by a procedure similar to that used above to define X_{21}.

Continue this process, finally constructing g_{2p_2}, X_{2p_2} and Y_{2p_2}. Let $D'_2 = g_{2p_2}(\Delta_0^2)$.

To construct D'_3, let $g_{30} = g_{2p_2}$, and consider the location of δ_{31} relative to X_{2p_2} and Y_{2p_2}. By following a procedure similar to that used above, we may construct $g_{31}, g_{32}, \cdots, g_{3p_3}, X_{31}, X_{32}, \cdots, X_{3p_3}, Y_{31}, Y_{32}, \cdots$, and Y_{3p_3}. We also construct $Z_{31}, Z_{32}, \cdots$, and Z_{3p_3} which are unions of certain ones of the δ_{3t}. Let $D'_3 = g_{3p_3}(\Delta_0^2)$.

Finally, to construct D', we let $g_{40} = g_{3p_3}$, and then construct $g_{41}, g_{42}, \cdots, g_{4p_4}$, X's, Y's, Z's, and a finite sequence $W_{41}, W_{42}, \cdots$, and W_{4p_4}, these being unions of certn ones of the δ_{4t}.

Let $g = g_{4p_4}$, and let $D' = g(\Delta_0^2)$. We shall show that D' has the properties required.

First, D' is an admissible modification of D relative to T_α. D' is constructed from D by a finite sequence of steps. In each such step, we make a replacement in some $T_{\alpha i}$, hence in T_α. Thus D' is admissible relative to T_α.

Let $X = X_{4p_4}$, $Y = Y_{4p_4}$, $Z = Z_{4p_4}$, and $W = W_{4p_4}$. Then clearly $g(X) \subset D_1 \cap T_{\alpha 1}$, $g(Y) \subset D_2 \cap T_{\alpha 2}$, $g(Z) \subset D_3 \cap T_{\alpha 3}$, and $g(W) \subset D_4 \cap T_{\alpha 4}$. Further, $g|\Delta_0^2 - (X \cup Y \cup Z \cup W)) = f|\Delta_0^2 - (X \cup Y \cup Z \cup W)$.

To show that $D' \cap T_{\alpha 1} \subset D_1$, note that $g^{-1}(T_{\alpha 1})$ and $Y \cup Z \cup W$ are disjoint. Further if $1 \leqq t \leqq p_1$ and δ_{1t} does not lie in X, then $\delta_{1t} \subset Y \cup Z \cup W$. From this and the fact that $g|\Delta_0^2 - (X \cup Y \cup Z \cup W) = f|\Delta_0^2 - (X \cup Y \cup Z \cup W)$, we get that on $(g^{-1}(T_{\alpha 1})) - X$, $g = f$ and so $g(g^{-1}(T_{\alpha 1}) - X)$ lies in D_1. Since $g(X) \subset D_1$, then $D' \cap T_{\alpha 1} \subset D_1$. Similarly we may show that if $i = 2, 3$, or 4, $D' \cap T_{\alpha i} \subset D_i$.

We may use the fact that $g|\Delta_0^2 - (X \cup Y \cup Z \cup W) = f|\Delta_0^2 - (X \cup Y \cup Z \cup W)$ to show that $D' - \cup_{i=1}^{4} T_{\alpha i} \subset D$. $\square$

The main step in our proof of the Generalized Bing Lemma is a generalization of Theorem 10 of [8]. In this step, we use the following idea from [8]. If β is any index and D and E are sets, then the core L_β of T_β has *Property P with respect to*

D and *E* if and only if there exist points x and y in opposite loops of L_β such that every arc in L_β from x to y intersects both D and E.

The next lemma provides the first step in the proof of the Generalized Bing Lemma. It follows easily from Lemma 4.2.

Lemma 4.3. *Suppose α is an index, and D and E are polyhedral singular discs in $\mathbb{R}^3$ such that (1) each of $Bd\,D$ and $Bd\,E$ is disjoint from T_α, (2) $D \cap E \cap T_\alpha = \emptyset$, and (3) each of D and E is in general position relative to $Bd\,T_\alpha$, and for $i = 1, 2, 3,$ or 4, to $Bd\,T_{\alpha i}$. Suppose that if $i = 1, 2, 3,$ or 4, there exist admissible modifications D_i and E_i of D and E, respectively, relative to $T_{\alpha i}$ such that (1) $D_i \cap E_i \cap T_{\alpha i} = \emptyset$ and (2) the core L_i of $T_{\alpha i}$ fails to have Property P with respect to D_i and E_i. Then there exist admissible modifications D' and E' of D and E, respectively, relative to T_α, such that (1) $D' \cap E' \cap T_\alpha = \emptyset$ and (2) if $i = 1, 2, 3,$ or 4, L_i fails to have Property P with respect to D' and E'.*

5. Trivial Curves of Intersection

In the main lemma of this paper, we have the following situation. For some index α, we have two polyhedral singular discs D and E with their boundaries outside T_α. For each i, $i = 1, 2, 3,$ or 4, we have a homotopy core L_i of $T_{\alpha i}$ such that L_i fails to have Property P with respect to D and E. Let $L = \cup_{i=1}^4 L_i$. We may assume that D and E are in general position relative to each of $Bd\,B_\alpha$ and $Bd\,C_\alpha$, and that neither D nor E intersects $L \cap (Bd\,B_\alpha \cup Bd\,C_\alpha)$.

Our goal in this situation will be to remove the intersection of D and E with each of B_α and C_α, preserving the failure of Property P.

In this section, we shall be concerned with curves of intersection of D and E with $Bd\,B_\alpha$ that are trivial on $(Bd\,B_\alpha) - L$, and with curves on $Bd\,C_\alpha$ with analogous properties. We shall show that we may remove such curves of intersection, with admissible modifications, preserving the failure of Property P.

Lemma 5.1. *Suppose α is an index, D and E are polyhedral singular discs such that (1) $Bd\,D$ and $Bd\,E$ are disjoint from T_α, (2) D and E are each in general position relative to $(Bd\,B_\alpha) \cup (Bd\,C_\alpha)$, and (3) $D \cap E \cap T_\alpha = \emptyset$. Suppose that if $i = 1, 2, 3,$ or 4, L_i is a homotopy core of $T_{\alpha i}$ such that L_i fails to have Property P with respect to D and E. Let $L = \cup_{i=1}^4 L_i$. Suppose that $D \cup E$ is disjoint from $L \cap (Bd\,B_\alpha \cup BD\,C_\alpha)$.*

Then there exist admissible modifications D' of D and E' of E, both relative to T_α, such that (1) $D' \cap E' \cap T_\alpha = \emptyset$, (2) if $i = 1, 2, 3,$ or 4, L_i does not have Property P with respect to D' and E', (3) each curve of intersection of D' and E' with $Bd\,B_\alpha$ is nontrivial on $(Bd\,B_\alpha) - L$, and (4) each curve of intersection of D' and E' with $Bd\,C_\alpha$ is nontrivial on $(Bd\,C_\alpha) - L$.

Proof. First we shall consider B_α. Let X be the universal covering space of $(Bd\,B_\alpha) - L$, and let $\pi : X \to (Bd\,B_\alpha) - L$ be projection. X is a plane, and if γ is a curve of intersection of D or of E with $Bd\,B_\alpha$ such that $\gamma \sim 0$ on $(Bd\,B_\alpha) - L$, then γ lifts to a loop $\tilde{\gamma}$ in X. If $\gamma \subset D$, let γ^* be the component of $\pi^{-1}(D)$ containing $\tilde{\gamma}$, and similarly if $\gamma \subset E$. By theorems of plane topology, there is a disc Γ_γ in X containing γ^* in its interior, and with its boundary close to γ^* and disjoint from both $\pi^{-1}(D)$ and $\pi^{-1}(E)$.

Since D and E each have only finitely many curves of intersection with $Bd\,B_\alpha$, it follows that there is a curve γ of intersection, either of D or of E with $Bd\,B_\alpha$, such that the disc Γ_γ intersects only one of $\pi^{-1}(D)$ and $\pi^{-1}(E)$, the one containing γ^*.

Suppose $\gamma \subset D$. Then $\widetilde{\gamma}$ bounds a singular disc Δ in Γ_γ disjoint from E. Then $\pi(\Delta)$ is a singular disc on $(Bd\,B_\alpha) - L$, bounded by γ, and disjoint from E. let Δ' be the subdisc of D bounded by γ. Replace Δ' by Δ, and adjust the resulting singular disc slightly to one side of $Bd\,B_\alpha$ to obtain an admissible modification of D with fewer curves of intersection with $Bd\,B_\alpha$ than D. A similar construction can be made if $\gamma \subset E$.

After finitely many repetitions of this procedure, we obtain modifications of D and E as required. A similar argument applies to C_α. $\square$

6. PRELIMINARIES

We turn now to establishing the main technical result for the proof of the Generalized Bing Lemma. This is Lemma 9.1 and its proof is given in sections 9, 10, and 11. This section and the next two are devoted to preliminary results for Lemma 9.1.

Throughout sections 6–11, we shall have the following standing hypotheses:

(1) α is an index.

(2) D and E are polyhedral singular discs in $\mathbb{R}^3$ such that (a) $Bd\,D \cup Bd\,E$ and T_α are disjoint, (b) both D and E are in general position relative to each of $Bd\,B_\alpha$ and $Bd\,C_\alpha$, and (c) $D \cap E \cap T_\alpha = \emptyset$.

(3) If $i = 1, 2, 3$, or 4, the core L_i of $T_{\alpha i}$ fails to have Property P with respect to D and E.

(4) (a) If γ is any curve of intersection of D with $Bd\,B_\alpha$, then $\gamma \not\sim 0$ on $(Bd\,B_\alpha) - \{x_1, x_2, x_3, x_4\}$. (b) If μ is any curve of intersection of E with $Bd\,B_\alpha$, then $\mu \not\sim 0$ on $(Bd\,B_\alpha) - \{x_1, x_2, x_3, x_4\}$. (c) Similar conditions hold relative to D, E, $Bd\,C_\alpha$, and the points of intersection of L_1, L_2, L_3, and L_4 with $Bd\,C_\alpha$. This last condition is justified by Lemma 5.1.

We shall abbreviate some of the notation in these six sections. By "modification" we shall mean "admissible modification relative to T_α", and by "Property $\overline{P}$" we shall mean "fails to have Property P with respect to D and E" or, if the context indicates, "fails to have Property P with respect to cetain modifications of D and E." The indices i, j, k, and ℓ are understood to take on values in the set $\{1, 2, 3, 4\}$ and, in any given context, are distinct.

If we modify the discs D and E to obtain D' and E', respectively, it is understood that (1) D' and E' are polyhderal singular discs in $\mathbb{R}^3$, and are modifications of D and E, respectively, (2) $D' \cap E' \cap T_\alpha = \emptyset$, (3) L_1, L_2, L_3, and L_4 have Property $\overline{P}$ with respect to D' and E', and (4) D' and E' satisfy appropriate general position assumptions.

If Δ is a polyhedral disc in $\mathbb{R}^3$, then the statement that we *modify D and E near* Δ means that we have modifications D' and E' of D and E, respectively, such that for some close neighborhood W of Δ, $D' \subset D \cup W$ and $E' \subset E \cup W$.

In these six sections, various constructions are made. We shall assume these made so that appropriate general position conditions are satisfied.

We shall use a technicque called *piping* to adjust D and E near the L's . Suppose A is an arc on some L_i, and D contains a point p of A. We replace a small disc of D containing p by a finger-shaped polyhedral disc extending over one end of A. We may make a similar construction for E. By making such constructions in a suitable order, we may preserve disjointness of D and E in T_α.

We apply this technique to the stems of L_1, L_2, L_3, and L_4. Thus in sections 6–11, we shall assume that neither D nor E intersects the stem of either L_1, L_2, L_3, or L_4.

Recall that the core L_α of T_α lies in a vertical plane π_α, and we assume that no three such planes intersect. If $i = 1, 2, 3$, or 4, we let $\Delta_{\alpha i} = B_\alpha \cap \pi_{\alpha i}$, $L_{\alpha i}^+ = L_{\alpha i} \cap B_\alpha$, and $L_{\alpha i}^- = L_{\alpha i} \cap C_\alpha$. The four cores $L_{\alpha 1}$, $L_{\alpha 2}$, $L_{\alpha 3}$, and $L_{\alpha 4}$ in T_α are placed slightly to some one side of the plane π_α. Let $\delta_{\alpha i}$ denote the disc in π_α bounded by the upper loop of $L_{\alpha i}$, and $\delta_{\alpha i}^-$ denote the disc in π_α bounded by the lower loop of $L_{\alpha i}$. Let X_i denote the point of $L_i \cap Bd\, B_\alpha$, and let y_i denote the point of $L_i \cap Bd\, C_\alpha$.

A D-cap in B_α is a polyhedral disc H spanning B_α such that (1) H is disjoint from E, (2) $Bd\, H$ contains no one of x_1, x_2, x_3, and x_4, but separates some two of them on $Bd\, B_\alpha$, and (3) for each i, $L_i^+ \cap H \subset D$. An E-cap in B_α is defined similarly, and we may define D-caps and E-caps for C_α.

Lemma 6.1. *If D intersects B_α, there is a D-cap either in B_α or C_α. Analogous results hold if either (a) D intersects C_α or (b) E intersects B_α or C_α.*

Proof. We shall consider the case where D intersects B_α. Let $B_\alpha' = \mathbb{R}^3 - \text{Int}\, B_\alpha$. Then there is a curve γ_0 of intersection of D with $Bd\, B_\alpha$ such that $\gamma_0 \nsim 0$ on $(Bd\, B_\alpha) = \{x_1, x_2, x_3, x_4\}$ and if d_0 is the subdisc of D bounded by γ_0, then Int d_0 is disjoint from $Bd\, B_\alpha$. Thus either $d_0 \subset B_\alpha$ or $d_0 \subset B_\alpha'$. If $d_0 \subset B_\alpha$, then by the Loop Theorem [16], there is a polyhedral disc H spanning B_α such that $H \cap E = \emptyset$, $Bd\, H \nsim 0$ on $(Bd\, B_\alpha) - \{x_1, x_2, x_3, x_4\}$, and H lies near d_0. Thus H is a D-cap in B_α.

If $d_0 \subset B_\alpha'$, then d_0 intersects C_α. If not, then by the Loop Theorem [15, 16], there is a polyhedral disc $\widehat{D}$ spanning B_α' such that $Bd\, \widehat{D} \nsim 0$ on $(Bd\, B_\alpha) - \{x_1, x_2, x_3, x_4\}$ with $\widehat{D}$ near d_0, and hence $\widehat{D} \cap C_\alpha = \emptyset$. But then $\widehat{D}$ intersects the stem of some L.

Thus there is a curve λ_0 of intersection of d_0 with $Bd\, C_\alpha$ such that $\lambda_0 \nsim 0$ on $(Bd\, C_\alpha) - (L_1 \cup L_2 \cup L_3 \cup L_4)$ and if d_1 is the subdisc of d_0 bounded by λ_0, then Int d_1 is disjoint from $Bd\, C_\alpha$. By the argument above, $d_1 \subset C_\alpha$. An application of the Loop Theorem [15, 16] to d_1 yields a D-cap in C_α. $\square$

Lemma 6.2. *Suppose $i \neq j$ and H is a D-cap in B_α such that $Bd\, H$ separates x_i from x_j on $Bd\, B_\alpha$. Then H intersect either L_i^+ or L_j^+. A similar result holds for L_i^-, L_j^-, and D-caps in C_α. Analogous results hold for E-caps in B_α or C_α.*

Proof. Suppose H intersects neither L_i^+ nor L_j^+. Let P be the disc on $Bd\, B_\alpha$ bounded by $Bd\, H$ and containing x_i, and let W^3 be the 3-cell in $\mathbb{R}^3$ bounded by $H \cup P$. Then $L_i^+ \subset W^3$ and since $Bd\, H$ separates x_i from x_j on $Bd\, B_\alpha$, $W^3 \cap L_j^+ = \emptyset$. Since $Bd\, \delta_i$ and $Bd\, \delta_j$ are linked, and $Bd\, \delta_i$ bounds a singular disc in W^3, it follows that L_j^+ intersects W^3. This is a contradiction. $\square$

The following is a key lemma of this paper. With its use, we may modify D and E, and preserve the fact that the L's have Property $\overline{P}$.

Lemma 6.3. *Suppose Δ is a polyhedral disc in Int T_α, and $J_1, J_2, \cdots,$ and J_n are mutually disjoint polyhedral simple closed curves in T_α, each intersecting Δ in* *exactly one point and piercing* Δ *there. Suppose that if* $m = 1, 2, \cdots,$ *or* n, J_m *intersects at most one of* D *or* E. *Suppose* W *is an open neighborhood of* Δ.

Suppose $D \cap Bd\,\Delta = \emptyset$. *Then there exist modifications* D' *and* E' *of* D *and* E, *respectively, such that (1)* $D' \cap \Delta = \emptyset$, *(2)* $D' \subset D \cup W$ *and* $E' \subset E \cup W$, *and (3)* D' *is disjoint from the same* J's *as* D, *and* E' *is disjoint from the same* J's *as* E.

Suppose $E \cap Bd\,\Delta = \emptyset$. *Then an analogous conclusion holds but with (1) above replaced by* $E' \cap \Delta = \emptyset$.

Proof. Suppose $D \cap Bd\,\Delta = \emptyset$, but $D \cap \Delta \neq \emptyset$. Note that each component of intersection of D with Δ is a curve of intersection. By the methods of proof of Lemma 5.1, there is a subdisc Δ_0 of Δ such that $Bd\,\Delta_0$ intersects neither D, E, nor $\overset{n}{\underset{m=1}{\cup}} J_m$, and Δ_0 intersects only one of D and E.

Suppose Δ_0 intersects D. Let J^* be the union of the J's that are disjoint from D. If each curve of intersection of D with Δ_0 is trivial on $\Delta_0 - J^*$, it is easy to modify D so as to eliminate such curves as curves of intersection. Now suppose that there is a curve of intersection γ of D with Δ_0 such that $\gamma \nsim 0$ on $\Delta_0 - J^*$. Then there is a curve of intersection γ_0 of D with Δ_0 such that (1) $\gamma_0 \nsim 0$ on $\Delta_0 - J^*$ and (2) if d is the subdisc of D bounded by γ_0 and λ is any curve of intersection of Int d with Δ_0, then $\lambda \sim 0$ on $\Delta_0 - J^*$. Clearly if λ is a curve as in (2) we may use a singular disc in $\Delta_0 - J^*$ bounded by λ to modify D and eliminate λ as a curve of intersection of D with Δ_0. Hence we may assume that Int d and Δ_0 are disjoint.

Construct a polyhedral 3-cell W^3 lying in a thin neighborhood of Δ_0, having Δ_0 on its boundary, and lying to the side of Δ_0 opposite that to which d abuts on Δ_0. Let $M^3 = [(\mathbb{R}^3 \cup \{\infty\}) - W^3] \cup \text{Int } \Delta_0$. By [15], it follows that $\gamma_0 \sim 0$ on $\Delta_0 - J^*$. This is a contradiction, and hence no such curve γ exists.

It follows that we may modify D so as to remove at least one curve of intersection of D with Δ, preserving disjointness of D and J^*. A similar argument applies if Δ_0 intersects E.

Since $D \cup E$ has only finitely many curves of intersection with Δ, then after finitely many repetitions of the process above, we will have modified D and E, obtaining D' and E' such that D' and E' have no curves of intersection with Δ. It follows that $D' \cap \Delta = \emptyset$.

If $E \cap Bd\,\Delta = \emptyset$, then an analogous argument applies. $\square$

Lemma 6.4. *Suppose the hypotheses of the first paragraph of Lemma 6.3. Suppose that* $1 \leq s < n$ *and if* $\leq t \leq s$, J_t *bounds a polyhedral disc* Δ_t *in* $\mathbb{R}^3$ *such that* $\Delta_t \cap \Delta$ *is an arc,* Δ_t *intersects at most one of* D *and* E, *and no three of* Δ, $\Delta_1, \Delta_2, \cdots,$ *and* Δ_s *intersect.*

If $D \cap Bd\,\Delta = \emptyset$, *then there exist modifications* D' *and* E' *of* D *and* E, *respectively, such that (1)* $D' \cap \Delta = \emptyset$, *(2)* $D' \subset D \cup W$ *and* $E' \subset E \cup W$, *and (3)* D' *is disjoint from the same* Δ_t's *and* J's *as* D, *and similarly for* E'.

If $E \cap Bd\,\Delta = \emptyset$, *and analogous conclusion holds with (1) replaced by (1)* $E' \cap \Delta = \emptyset$.

Proof. Suppose $D \cap Bd\,\Delta = \emptyset$. Let $t_1, t_2, \cdots,$ and t_r be the indexes t such that Δ_t is disjoint from D. Then $(\text{Int } \Delta) = \overset{r}{\underset{p=1}{\cup}} \Delta_{t_p}$ is an open disc containing $D \cap \Delta$. Similarly, if $s_1, s_2, \cdots,$ and s_u are the indexes s such that $E \cap \Delta_s = \emptyset$, then $(\text{Int } \Delta) - \overset{u}{\underset{q=1}{\cup}} \Delta_{s_q}$ is an open disc containing $E \cap \Delta$. Then Lemma 6.4 follows from Lemma 6.3. $\square$

Lemma 6.5. *Suppose H is a D-cap in B_α and $D \cap Bd\,\delta_i = H \cap Bd\,\delta_i = \emptyset$. Then there exist a D-cap H' in B_α and modifications D' and E' of D and E, respectively, near δ_i, such that $Bd\,H' = Bd\,H$, $H' \cap \delta_i = \emptyset$, and each of L_1, L_2, L_3, and L_4 has Property $\overline{P}$ with respect to D' and E'.*

Proof. Each component of $\delta_i \cap H$ is a simple closed curve in Int δ_i. Let d be a disc on H bounded by such a curve and with $(\text{Int } d) \cap H = \emptyset$. Suppose $j \neq i$. We may, by piping D and E near $Bd\,\delta_j$, assume that if $Bd\,\delta_j$ intersects d in p_j, then the component ξ_j of $(Bd\,\delta_j) - H$ containing p_j intersects at most one of D and E. If $d \cap Bd\,\delta_j = \emptyset$, let $\xi_j = \emptyset$. In $\mathbb{R}^3 \cup \{\infty\}$, we may let H be the point at infinity. Then Int d is a plane and if $d \cap Bd\,\delta_j \neq \emptyset$, ξ_j is a line. By an argument similar to that given for Lemma 6.4, we may assume that $d \cap E = \emptyset$.

Let d' be the subdisc of H bounded by $Bd\,d$. We replace d' by a close copy of d, and adjust so as to reduce the number of curves of intersection of H with δ_i. After finitely many steps, we obtain a D-cap H' in B_α as required. $\square$

7. FINS

Fins are polyhedral discs in B_α and C_α that we shall use to simplify the intersections of the polyhedral singular discs D and E with B_α and C_α. In this section, we shall concentrate on B_α, but analogous results hold for C_α.

A *fin* or *half-fin* in B_α is a polyhedral disc F in B_α such that $F \cap Bd\,B_\alpha = (Bd\,F) \cap Bd\,B_\alpha$ is an arc σ, the *base* of F. Thus $F - \sigma \subset \text{Int } B_\alpha$; the arc $(BdF) - (\text{Int } \sigma)$ is the *edge* of F.

Recall that in sections 2 and 6, for $i = 1, 2, 3$, and 4, we described planes $\Pi_{\alpha i}$ in $\mathbb{R}^3$ and discs $\Delta_{\alpha i} = B_{\alpha i} \cap \Pi_{\alpha i}$ in $B_{\alpha i}$. In sections 7–11, let Δ_i denote $\Delta_{\alpha i}$.

Fins in B_α are constructed from two Δ's. Suppose $i \neq j$. There are three components of $\Delta_i - [L_i^+ \cup \delta_i \cup (\Delta_i \cap \Delta_j)]$. There are four fins in B_α constructed from these sets; see Figure 3. The base of each such fin is an arc on $Bd\,B_\alpha$ from x_i to x_j, and the edge of such a fin lies in $L_i^+ \cup L_j^+ \cup (\delta_i \cap \delta_j)$. These fins are called *fins in B_α for L_i and L_j*.

Suppose F is such a fin in B_α, with base σ. F is disjoint from $(\text{Int } \delta_i) \cup (\text{Int } \delta_j)$, and note that (1) there is an open neighborhood of $(\text{Int } F) \cup \sigma$ disjoint from $[(Bd\,\delta_i) \cup (Bd\,\delta_j)] - F$, and (2) if $k \neq i$ and $k \neq j$, then $Bd\,\delta_k$ contains at most one point of F, and $F \cap \delta_k$ is $\emptyset$ or an arc with exactly one point, an end point, on $Bd\,F$. For these reasons, we say F is *good*.

In a number of cases, we have a polygonal arc λ on $Bd\,B_\alpha$, and construct a fin F in B_α with base σ whose endpoints are not on λ. Then by adjusting F near λ, we may assume that $\sigma \cap \lambda = \emptyset$.

A *half-fin* in B_α is constructed using just one disc Δ, and requires a D-cap or an E-cap in B_α. Suppose $i = 1, 2, 3$, or 4 and X is either a D-cap or an E-cap in B_α such that X separates x_i and some point of $Bd\,\delta_i$ in B_α. Let A^* be the union of all arc components of $X \cap (\Delta_i - \text{Int } \delta_i)$ that have one endpoint on $Bd\,B_\alpha$ and the other on L_i^+, hence on $Bd\,\delta_i^+$. There are exactly two components of $(\Delta_i - \text{Int } \delta_i) - A^*$ that have x_i as a limit point. The closure of each of these is a *half-fin for L_i* and X. Note that the base of this half-fin is an arc on $Bd\,B_\alpha$ with x_i as one endpoint.

If $i \neq j$, a *slit on $Bd\,B_\alpha$ from x_i to x_j* is a polygonal arc on $Bd\,B_\alpha$ with endpoints x_i and x_j, and containing no other of the x's. If Γ is a polygonal simple closed curve on $Bd\,B_\alpha$ containing no x, then a *cut on $Bd\,B_\alpha$ from x_i to Γ* is a polygonal arc σ on $Bd\,B_\alpha$ from x_i to a point of Γ, containing no other x's, and with $\Gamma \cap (\text{Int } \sigma) = \emptyset$.

Lemma 7.1. *Suppose i, j, and k are distinct, and F_j and F_k are fins in B_α using L_i and L_j, and L_i and L_k, respectively. Then we may adjust F_j and F_k to obtain fins F_j' and F_k' in B_α such that (1) $F_j' \cap F_k' = \{x_i\}$, (2) the base of F_j' is a slit on $Bd B_\alpha$ from x_i to x_j, (3) the base of F_k' is a slit on $Bd B_\alpha$ from X_i to X_k, and (4) the edges of F_j' and F_k' are arbitrarily close to those of F_j and F_k, respectively.*

Proof. The fins F_j and F_k may have in common a subdisc of δ_i used to construct these fins; see the description above. We may adjust one of the F's slightly, keeping the edge pointwise fixed, to eliminate such a common disc. F_j and F_k may also have in common one or more arcs not on the edge of either; see Figure 4. Such arcs of intersection may be removed as indicated in Figure 4. These procedures yield fins whose common part lies in the edge of each. An arbitrarily small adjustment, keeping x_i fixed, will eliminate such intersections. $\square$

In the next lemma, we shall use a procedure called *creasing* to remove certain arcs of intersection of a polyhedral disc and a polyhedral singular disc.

Lemma 7.2. *Suppose F is a fin or half-fin in B_α with edge e and base σ. Suppose W is a neighborhood of $\sigma \cup Int\, F$. Suppose that $D \cap e = \emptyset$. Then there exist modifications D' and E' of D and E, respectively, such that (1) $D \cap F = \emptyset$, and (2) $D' \subset D \cup W$ and $E' \subset E \cup W$. An analogous result holds if $E \cap e = \emptyset$.*

Proof. By using ideas of the proof of Lemma 5.1, we may assume that neither D nor E has any curve of intersection with F. Suppose β is an arc of intersection of D with F. Then β is a singular arc in F with both endpoints on σ. Suppose Λ is a component of $F \cap D$. Then there is a polyhedral subdisc Λ^* of F such that (1) $Bd\,\Lambda^*$ is the union of arcs λ_Λ spanning F and disjoint from e, and ζ_Λ lying in $Int\, \sigma$, (2) λ_Λ is near Λ^* and hence disjoint from E, and (3) $\Lambda \subset \zeta_\Lambda \cup Int\, \Lambda^*$. The discs Λ^* constructed in this way contain $F \cap D$. Any two such discs are either disjoint or one contains the other.

The discs Λ^* may intersect E. Any such point of E lies on an arc of intersection of E with F, having both endpoints on ζ_Λ. If Λ^* is a disc as above, and $E \cap \Lambda^* \neq \emptyset$, let Γ be a component of $E \cap \Lambda^*$, and make a construction similar to that above to obtain an analogous polyhedral subdisc Γ^* of Λ^* related to E as Λ^* is to D.

We may continue this process, defining discs in F relative to D and E alternately. This process terminates after finitely many steps.

Let Ω^* be one of the subdiscs of F constructed at the last step of this process. Then Ω^* intersects only of D and E. Suppose $\Omega^* \cap D \neq \emptyset$. Then $Bd\,\Omega^* = \omega \cup \omega'$ where ω is a spanning arc of F disjoint from D and E and $\omega' \subset \sigma$. Each point of $\Omega^* \cap D$ lies on an arc of intersection of D and F having both endpoints on ω'.

Suppose β is such an arc. Let β' be the subarc of ω' with the same endpoints at β. Then $\beta \cup \beta'$ bounds a singular disc $\hat{\beta}$ in Ω^*. Using a doubled and folded copy of $\hat{\beta}$, we may "crease" D and remove β as an arc of intersection of D with F. We may thus adjust D near Ω^* so as to remove all points of intersection of D with Ω^*. A similar argument applies if $\Omega^* \cap E \neq \emptyset$.

After finitely many repetitions of this procedure, we obtain D' and E' as required. $\square$

Lemma 7.3. *Suppose the hypotheses of the first paragraph of Lemma 6.3. Suppose that if $1 \leq t \leq n$, J_t bounds a polyhedral disc Δ_t in $\mathbb{R}^3$ such that $\Delta_t \cap \Delta$ is an arc, Δ_t intersects at most one of D and E, and no three of $\Delta, \Delta_1, \Delta_2, \cdots$, and*

Δ_n intersect. Suppose β is an arc on $Bd\,\Delta$, $\zeta = (Bd\,\Delta) - Int\,\beta$, and for each t, $q \leqq t \leqq n$, $\Delta_t \cap Bd\,\Delta \subset Int\,\zeta$.

Suppose $D \cap \zeta = \emptyset$. Then there exist modifications D' and E' of D and E, respectively, such that (1) $D' \cap \Delta = \emptyset$, (2) $D' \subset D \cup W$ and $E' \subset E \cup W$, and (3) D' is disjoint from the same Δ_t's as D, and E' is disjoint from the same Δ_t's as E.

If $E \cap \zeta = \emptyset$, then an analogous conclusion holds but with (1) replaced by (1) $E' \cap \Delta = \emptyset$.

Proof. Suppose $D \cap \zeta = \emptyset$. Let $t_1, t_2, \cdots$, and t_m be the indexes t such that Δ_t is disjoint from D. Then $(Int\,\Delta) - (\overset{m}{\underset{r=1}{\cup}} \Delta_{t_r})$ is an open disc containing $D \cap \Delta$. Similarly, if $s_1, s_2, \cdots$, and s_u are the indexes s such that Δ_s is disjoint from E, then $(Int\,\Delta) - (\overset{u}{\underset{q=1}{\cup}} \Delta_{s_q})$ is an open disc containing $E \cap A$. Then Lemma 7.3 follows from the proof of Lemma 7.2. $\square$

8. The S-cases

We have an *S-case in B_α* if there exists a polygonal simple closed curve S on $Bd\,B_\alpha$ such that

(1) $(Bd\,B_\alpha) - S$ is the union of disjoint connected open sets $\widehat{U}$ containing x_1 and $\widehat{V}$ containing x_4, $L_1 \cap D = \emptyset$, and $L_4 \cap E = \emptyset$,

(2) there is a cut τ in $\widehat{U}$ from x_1 to a point of S such that if $x_i \in \widehat{U}$, then $x_i \in \tau$ and $L_i^- \cap D = \emptyset$, and

(3) there is a cut σ in $\widehat{V}$ from x_4 to a point of S such that if $x_i \in \widehat{U}$, then $x_i \in \sigma$ and $L_i^- \cap E = \emptyset$.

We may have an S-case in B_α with one or both of x_2 and x_3 on S. We describe these as an *S-case in B_α with $x_2 \in S$*, or an *S-case in B_α with $x_3 \in S$*, or an *S-case in B_α with both x_2 and x_3 on S*.

Analogous definitions hold for C_α.

Our first objective is to show that we have an S-case in B_α provided the following, *Hypothesis A*, holds: Suppose that there exist disjoint polygonal arcs τ and σ on $Bd\,B_\alpha$ such that

(1) each x_t lies on one of σ and τ,

(2) $x_1 \in \tau$, $\tau \cap D = \emptyset$, $L_1 \cap D = \emptyset$, and if $x_t \in \tau$, $L_t^- \cap D = \emptyset$, and

(3) $x_4 \in \sigma$, $\sigma \cap E = \emptyset$, $L_4 \cap E = \emptyset$, and if $x_t \in \sigma$, $L_t^- \cap E = \emptyset$.

Lemma 8.1. *Suppose Hypothesis A holds. There exist a fin F in B_α for L_i and L_j and a point q of the edge e of F such that*

(1) *the subarc e' of e from x_i to q is disjoint from D and*

(2) *the subarc e'' of e from q to x_j is disjoint from E.*

Proof. Since $L_i \cap D = \emptyset$ and $L_j \cap E = \emptyset$, then by Lemma 6.4, we may assume D and E modified near δ_i so that $D \cap \delta_i = \emptyset$ and $L_j \cap E = \emptyset$. Then there is a fin F in B_α for L_i and L_j and whose edge lies in $L_i^+ \cup (\delta_i \cap \delta_j) \cup L_j^+$. Let q be the point of $\delta_i \cap Bd\,\delta_j$ on the edge e of F. The subarc $x_i q$ of e lies in $L_i^+ \cup \delta_i$, and hence is disjoint from D. The subarc $q z_j$ of e lies in L_j, and hence is disjoint from E. $\square$

Lemma 8.2. *Suppose the hypothesis and notation of Lemma 8.1. Then there is a polygonal arc qr spanning F from q to a point r of the base of F such that the arc qr is disjoint from $D \cup E$.*

Proof. Let E^* be the union of all components of $E \cap F$ intersecting e'. Let D^* be the union of all components of $D \cap F$ intersecting e''. Let u and v be the last point of E^* and the first point of D^*, respectively, on the base β of F in the order from x_i to x_j on β. We claim that u precedes v on β in this order. If not, then by theorems of plane topology, D and E intersect. It then follows that there is a polygonal arc spanning F from q to some point r of β, lying near E but disjoint from $D \cap E$. $\square$

Lemma 8.3. *Assume the hypothesis and notation of Lemma 8.2. We may assume that if β' and β'' are the subarcs $x_i r$ and rx_j, respectively, of β, then $D \cap \beta' = \emptyset$ and $E \cap \beta'' = \emptyset$.*

Proof. This follows from Lemma 7.2 applied twice. $\square$

Lemma 8.4. *Suppose the hypothesis and notation of Lemma 8.3. Then there exists a polyhedral simple closed curve S on $Bd\,B_\alpha$ such that*

(1) *S separates x_i and x_j on $Bd\,B_\alpha$,*
(2) *S is disjoint from D, E, σ, and τ, and*
(3) *S intersects β and in precisely the point r.*

Proof. Since the edge e of R is an unknotted spanning arc of B_α, then we may assume that the base β of F intersects τ only in x_i and σ only in x_j.

Let E^* be the union of $\tau \cup \beta'$ and all components of $E \cap Bd\,B_\alpha$ that intersect $\tau \cup \beta'$. This is a compact connected set on $Bd\,B_\alpha$, disjoint from D, and intersecting $\sigma \cup \beta''$ in only one point r. By theorems of plane topology, there is a polygonal simple closed curve S on $Bd\,B_\alpha$ containing r, disjoint from D, E, $\tau \cup (\beta' - \{r\})$, and $\sigma \cup (\beta'' - \{r\})$, and separating E^* from $\sigma \cup (\beta'' - \{r\})$ on $Bd\,B_\alpha$. It follows that S separates x_i from x_j on $Bd\,B_\alpha$, and itnersects β precisely in the point r. $\square$

Lemma 8.5. *If Hypothesis A holds, then we have an S-case in B_α.*

Lemma 8.6. *Suppose that we have an S-case in B_α, and W is a neighborhood of B_α. Then there exist modifications D' and E' of D and E, respectively, such that*

(1) *$(D' \cup E') \cap B_\alpha = \emptyset$,*
(2) *$D' \subset D \cup W$ and $E' \subset E \cup W$, and*
(3) *each L_t has Property $\overline{P}$ with respect to D' and E'.*

Proof. We shall use the notation of the preceding part of this section. Let Σ_1 be the disc on $Bd\,B_\alpha$ bounded by S and containing x_1. We claim that $\Sigma_1 \cap D = \emptyset$. Otherwise, there is a curve of intersection γ of D in Σ_1. Since $\beta' \cap D = \emptyset$, γ lies in the open disc $\Sigma_1 - \beta'$. Then $\gamma \sim 0$ on $(Bd\,B_\alpha) - \{x_1, x_2, x_3, x_4\}$, and this is a contradiction.

Let Σ_2 be the disc on $Bd\,B_\alpha$ bounded by S and containing x_4. It is clear that $\Sigma_2 \cap E = \emptyset$.

It is now straightforward to modify D and E to obtain D' and E' as required. $\square$

Let *Hypothesis B* denote the following: There exist polyhedral arcs τ, σ_3, and σ_4 on $Bd\,B_\alpha$ with endpoints x_1 and x_2, x_2 and x_3, and x_2 and x_4, respectively, such that

(1) *$\tau \cap (\sigma_3 \cap \sigma_4) = \emptyset$ and $\sigma_3 \cap \sigma_4 = \{x_2\}$,*

(2) $\tau \cap D = \emptyset$, and $(\sigma_3 \cup \sigma_4) \cap E = \emptyset$, and

(3) $L_1 \cap D = \emptyset$, $L_4 \cap E = \emptyset$, and $L_e^- \cap E = \emptyset$.

Lemma 8.7. *If Hypothesis B holds we have an S-case in B_α with $x_2 \in S$.*

Lemma 8.8. *Suppose that Hypothesis A holds and in addition, there is a slit ω in $Bd\,B_\alpha$ from x_2 to x_3 with $Int\,\omega$ disjoint from $\sigma \cup \tau$, and $\omega \cap (D \cup E) = \emptyset$. Then we have an S-case in B_α with both x_2 and x_3 on S.*

Lemma 8.9. *Suppose that we have an S-case in B_α, with one or both of x_2 and x_3 on S, and W is an open neighborhood of B_α. Then there eixst modifications D' and E' of D and E, respectively, such that*

(1) $(D' \cup E') \cap B_\alpha = \emptyset$,

(2) $D' \subset D \cup W$ and $E' \subset E \cup W$, and

(3) *each L_t has Property $\overline{P}$ with respect to D' and E'.*

9. MODIFYING SINGULAR DISCS

Lemma 9.1. *Suppose that α is an index, and D and E are polyhedral singular discs in $\mathbb{R}^3$ such that*

(1) $Bd\,D$ *and* $Bd\,E$ *are disjoint from* T_α,

(2) $D \cap E \cap T_\alpha = \emptyset$, *and*

(3) *if $i = 1, 2, 3$, or 4, the core L_i of $T_{\alpha i}$ has Property $\overline{P}$ with respect to D and E.*

Suppose that U is a neighborhood of $Bd\,B_\alpha$. Then there exist admissible modifications D' and E' of D and E, respectively, relative to T_α, such that

(1) $D' \cap E' \cap T_\alpha = \emptyset$,

(2) $(D' \cup E') \cap B_\alpha = \emptyset$,

(3) $D' \subset D \cup U$ and $E' \subset E \cup U$, and

(4) *if $i = 1, 2, 3$, or 4, then L_i has Property $\overline{P}$ with respect to D' and E'.*

Proof. Recall the standing hypotheses stated at the beginning of section 6. In particular, we assume that neither D nor E intersects the stem of any L.

In this proof, discs designated as D or E are assumed not to intersect the stem of any L, but discs designated as D' or E' may intersect stems of the L's.

Case 0.   Suppose that $D \cap B_\alpha = \emptyset$. If also $E \cap B_\alpha = \emptyset$, the lemma holds, so we suppose that $E \cap B_\alpha \neq \emptyset$. First suppose that D is disjoint from L_2, L_3, and L_4. Since every curve of intersection of E with $Bd\,B_\alpha$ is trivial on $(Bd\,B_\alpha) - \{x_1\}$, we may modify E near $Bd\,B_\alpha$ to obtain E' such that $E' \cap L_1^+ = \emptyset$. Let $D' = D$. Then D' is disjoint from L_2, L_3, and L_4, and L_1 has Property $\overline{P}$ with respect to D' and E'.

Suppose D intersects L_1 and some other L. Then $E \cap L_1^+ = \emptyset$ and by Lemma 6.4, we may assume in addition that $E \cap \delta_1 = \emptyset$.

Suppose for example that D intersects each L. Then E is disjoint from each L^+. By using the bottom edges of the δ's, we may construct fins in B_α, Σ_2 for L_1 and L_2, Σ_3 for L_1 and L_3, and Σ_4 for L_1 and L_4, such that E intersects the edge of no Σ and the bases of any two Σ's have only x_1 in common. We may adjust the Σ's slightly so that Σ_2 is disjoint from the edge of Σ_3, and both Σ_2 and Σ_3 are disjoint from the edge of Σ_4. If $i = 1, 2, 3$ or 4, let σ_i be the base of Σ_i.

We first modify E near Σ_2, using Lemma 7.2, and may assume that $\sigma_2 \cap E = \emptyset$; note that E is still disjoint from the edges of Σ_3 and Σ_4. We next modify E near E_3, using Lemma 7.2 and may assume that $E \cap \sigma_3 = \emptyset$; E is still disjoint from the edge of Σ_4 and from σ_2. Finally we modify E near Σ_4, using Lemma 7.2, and may assume $E \cap \sigma_4 = E \cap \sigma_3 = E \cap \sigma_2 = \emptyset$.

Let $\widehat{\sigma} = \sigma_2 \cup \sigma_3 \cup \sigma_4$; $\widehat{\sigma}$ is a triod. Each curve of intersection of E with $Bd\, B_\alpha$ is trivial on $(Bd\, B_\alpha) - \widehat{\sigma}$. Hence by Lemma 5.1, there is a modification E' of E such that $E' \cap B_\alpha = \emptyset$ and for each i, $E' \cap L_i \subset E \cap L_i$. Let $D' = D$.

Suppose now that L_1 and L_2 intersect D but not E, and L_3 and L_4 intersect E but not D. By Lemma 6.4 we may assume that $\delta_1 \cap E = \delta_2 \cap E = \emptyset$ and $\delta_3 \cap D = \delta_4 \cap D = \emptyset$. Let F_1 be a fin in B_α for L_1 and L_2, with base σ and with edge disjoint from E. By Lemma 7.3, we may assume that, in addition, $\sigma \cap E = \emptyset$. Let F_2 be a fin in B_α for L_3 and L_4 with base τ and edge disjoint from D; we may assume $\tau \cap \sigma = \emptyset$. By Lemma 7.3, we may assume that $\tau \cap D = \emptyset$. We have an S-case in B_α, and may apply Lemma 8.6.

If D and E both intersect L_1, then $L_1^+ \cap (D \cup E) = \emptyset$. Thus if also L_2 intersects D but not E, and both L_3 and L_4 intersect E but not D, we may assume that we have an S-case in B_α with $x_1 \in S$, and may apply Lemma 8.9.

Other cases where $D \cap B_\alpha = \emptyset$ may be treated in a similar manner. Thus we may assume that $D \cap B_\alpha \neq \emptyset$, and by a similar argument, that $E \cap B_\alpha \neq \emptyset$.

10. Case I

By Lemma 6.1, there is either a D-cap or an E-cap either in B_α or in C_α. We *first consider the case where there is a D-cap H in B_α.* Then $(Bd\, H) \sim 0$ on $(Bd\, B_\alpha) = \{x_1, x_2, x_3, x_4\}$. It follows that $Bd\, H$ separates some two x's on $Bd\, B_\alpha$.

Let P be the disc on $Bd\, B_\alpha$ bounded by $Bd\, H$ and containing x_1, and let $\Omega = (Bd\, B_\alpha) - (\text{Int } P)$. Let W^3 and X^3 be the 3-cells in B_α bounded by $H \cup P$ and $H \cup \Omega$, respectively.

<u>Case</u> I. $Bd\, H$ separates x_1 and x_2 from x_3 and x_4 on $Bd\, B_\alpha$.

<u>Case</u> IA. $L_1^+ \cap H = \emptyset$ and $L_2^+ \cap H \neq \emptyset$.

It follows that $E \cap \Omega = \emptyset$. Otherwise there is an E-cap K with base Q, $Q \subset \text{Int } \Omega$, and K either in X^3 or in $C\ell(\mathbb{R}^3 - X^3)$. Since L_3 and L_4 have Property $\overline{P}$, then K cannot lie in X^3. If K lies in $C\ell(\mathbb{R}^3 - X^3)$, we have a contradiction similar to that of the proof of Lemma 6.2. It follows that $E \cap X^3 = \emptyset$, and hence L_3 and L_4 are disjoint form E.

If $E \cap B_\alpha = \emptyset$, we have Case 0. Then $E \cap B_\alpha \neq \emptyset$ and by Lemma 6.1, there is E-cap in B_α or in C_α.

Suppose first there is an E-cap K in B_α, with base Q. Clearly $Q \subset P$, $x_1 \in \text{Int } Q$, and $K \subset W^3$. Using Δ_1, we may construct a half-fin F_1 in B_α with edge lying in $L_1^+ \cup K$, and disjoint from D. We may assume that F_1 lies in Y^3, the 3-cell in B_α bounded by $K \cup Q$. If τ is the base of F_1, τ is a cut on $Bd\, B_\alpha$ from x_1 to $Bd\, K$. By Lemma 7.2, we may assume in addition that $D \cap \tau = \emptyset$; the modifications made here are made near F_1. Then by Lemma 5.1, we may assume that $D \cap Q = \emptyset$. In this case, $L_1 \cap D = \emptyset$.

Suppose there is no E-cap in B_α. Then by Lemma 6.1, there is an E-cap K in C_α. Clearly the base Q of K contains y_1 and none of y_2, y_3, or y_4. Further, K intersects L_1^-. By an argument similar to that above, we may assume that $D \cap L_1 = \emptyset$.

We shall show now that we may assume that $C_\alpha \cap (D \cup E) = \emptyset$. Since $L_1^- \cap D = \emptyset$ and $L_2^- \cap E = L_3^- \cap E = L_4^- \cap E = \emptyset$, we may assume by Lemma 6.4 that $\delta_1^- \cap D = \emptyset$ and $\delta_2^- \cap E = \delta_3^- \cap E = \delta_4^- \cap E = \emptyset$. There is a fin F_3 in C_α for L_2 and L_3 with edge disjoint from E and base a slit σ_{23} in $Bd\,C_\alpha$ from y_2 to y_3. By Lemma 7.2, we may assume $E \cap \sigma_{23} = \emptyset$. There is a fin F_4 in C_α for L_2 and L_4 with edge disjoint from E and base a slit σ_{24} in $Bd\,C_\alpha$ from y_2 to y_4. Since the edge of F_4 is an unknotted spanning arc of C_α, we may assume that $\sigma_{23} \cap \sigma_{24} = \{y_2\}$. We may then assume by Lemma 7.2 that also $\sigma_{24} \cap E = \emptyset$. Then if $L_2^+ \cap E = \emptyset$, we have an S-case in C_α.

If L_2^+ intersects E, we may use a fin in C_α for L_1 and L_2 to construct a slit τ_{12} in $Bd\,C_\alpha$ from y_1 to y_2, and may assume that $D \cap \tau_{12} = \emptyset$. In this case, we have an S-case in C_α with $y_2 \in S$. Either by Lemma 8.6 or Lemma 8.9, we may assume that D and E are disjoint from C_α. We may also assume, by piping, that D and E intersect no stem of any L.

We may assume that if μ is any curve of intersection of E with P, then $\mu \sim 0$ in $P - (\{x_2\} \cup D)$. Suppose not. Let β be an arc in L_2^+ from x_2 to a point of H. There is an E-cap K' for W^3 with base Q' in P such that $(Bd\,K') \nsim 0$ in $P - (\{x_2\} \cup D)$. Clearly $x_2 \notin Q'$, and hence $x_1 \in Q'$. Since $E \cap C_\alpha = \emptyset$, then $K' \subset W^3$. By Lemma 6.2, L_1^+ intersects K'. Since $(Bd\,K') \nsim 0$ in $P - \{x_2\} \cup D$, then Q' contains a curve of intersection of D with Q'. By an argument as above, we may modify D and E so as to remove any such curve of interseciton.

Since D has only finitely many curves of intersection with $Bd\,B_\alpha$, this process termiantes after finitely many steps. By Lemma 5.1, we may then assume $E \cap B_\alpha = \emptyset$. This is Case 0.

Case IB.   Neither L_1^+ nor L_2^+ intersects H.

Case IB1.   There is an E-cap K in W^3 with base Q in P and containing both x_1 and x_2.

By Lemma 6.2, both L_1^+ and L_2^+ intersect K. We shall now adjust H, K, D, and E. By Lemma 6.5, we may assume H is disjoint from δ_1. Since $\delta_1 \cap \delta_3$ is an arc with one endpoint on $Bd\,\Delta_3$, then by Lemma 6.5, we may assume that, in addition, K is disjoint from δ_3. Since $\delta_1 \cap \delta_4$ and $\delta_3 \cap \delta_4$ are disjoint arcs, each with an endpoint on $Bd\,\delta_4$, by Lemma 6.5, we may assume that also $K \cap \delta_4 = \emptyset$.

There is a subdisc ϵ_1 of δ_1 such that $Bd\,\epsilon_1$ is disjoint from D and $\delta_1 \cap \delta_3 \subset \epsilon_1$. By Lemma 6.4, we may assume that $\epsilon_1 \cap D = \emptyset$; these modifications are made near ϵ_1. Thus there is a fin F_3' in B_α for L_1 and L_3 such that the edge of F_3' is the union of two arcs ζ_3 and ζ_3' where

$$x_1 \in \zeta_3, \ x_3 \in \zeta_3', \ \zeta_3 \cap D = \emptyset, \text{ and } \zeta_3' \cap E = \emptyset. \tag{$*$}$$

By a similar argument, there is a fin F_4' and B_α for L_1 and L_4 such that the edge of F_4' is the union of two arcs ζ_4 and ζ_4' where

$$x_1 \in \zeta_4, \ x_4 \in \zeta_4', \ \zeta_4 \cap D = \emptyset, \text{ and } \zeta_4' \cap E = \emptyset. \tag{$**$}$$

By Lemma 7.1, we may adjust F_3' and F_4' to obtain fins F_3'' and F_4'' in B_α such that

(1) $F_3'' \cap F_4'' = \{x_1\}$,
(2) $x_3 \in F_3''$ and $x_4 \in F_4''$, and
(3) the edge of F_3'' has property $(*)$ and the edge of F_4'' has property $(**)$.

Let $\widehat{F}_3$ be the component of $F_3'' \cap X^3$ containing x_3, and let $\widehat{F}_4$ be the component of $F_4'' \cap X^3$ containing x_4. By adjusting these, we may obtain disjoint half fins in B_α, F_3 for L_3 and H, and F_4 for L_4 and H, each having its edge disjoint from E.

Thus there exist disjoint cuts in Ω, σ_3 from $Bd\,H$ to x_3 and σ_4 from $Bd\,H$ to x_4, and by Lemma 7.2, we may assume $\sigma_3 \cap E = \sigma_4 \cap E = \emptyset$. By Lemma 5.1, we may assume $\Omega \cap E = \emptyset$. Then $L_3 \cap E = L_4 \cap E = \emptyset$. Note that the modifications occur on a close neighborhood of X^3.

By a similar construction involving L_1, L_2, L_4, and K, we may assume that $Q \cap D = \emptyset$ and $L_1 \cap D = L_2 \cap D = \emptyset$.

We now have an S-case in B_α.

Case IB2.   There is an E-cap K in C_α with base Q containing both y_1 and y_2.

It follows from Lemma 6.2 that (1) neither y_3 nor y_4 lies in Q, and (2) both L_1^- and L_2^- intersect K.

Since $L_1^+ \cap D = L_2^+ \cap D = \emptyset$, then by an argument similar to one used in Case IB1, we may assume that $E \cap \Omega = \emptyset$. Thus $L_3 \cap E = L_4 \cap E = \emptyset$. By an argument similar to that of Case IB1, we may also assume that $Q \cap D = \emptyset$.

By Lemma 6.4, we may assume that $\delta_1 \cap D = \delta_2 \cap D = \emptyset$ and $\delta_3 \cap E = \delta_4 \cap E = \emptyset$. There is a fin F_{12} in B_α for L_1 and L_2, with edge disjoint from D. Thus if τ_{12} is the base of F_{12}, τ_{12} is a slit in $Bd\,B_\alpha$, and by Lemma 7.2, we may assume that, in addition, $\tau_{12} \cap D = \emptyset$. We now have an S-case in B_α.

Case IB3.   There exist (1) an E-cap K_1 with base Q_1 either (a) in B_α, with $x_1 \in Q_1$ but $x_2 \notin Q_1$, or (b) in C_α, with $y_1 \in Q_1$ but $y_2 \notin Q_1$, and 92) an E-cap K_2 with base Q_2 either (a) in B_α, with $x_2 \in Q_2$ but $x_1 \notin Q_2$, or (b) in C_α, with $y_2 \in Q_2$ but $y_1 \notin Q_2$.

By arguments similar to some used in Case IA, we may assume that $L_1 \cap D = L_2 \cap D = \emptyset$. Then by an argument similar to one of Case IB1, we may assume that $\Omega \cap E = \emptyset$. This case is now similar to Case IB2.

Case IB4.   There is an E-cap K with base Q, either (a) in B_α with $x_1 \in Q$ or (b) in C_α with $y_1 \in Q$, but no E-cap K' with base Q', either (a) in B_α with $x_2 \in Q'$ or (b) in C_α with $y_2 \in Q'$.

We may, in either case, assume that, additionally, $L_1 \cap D = \emptyset$. If either D or E intersects L_2^-, then by an argument like one used in Case IB1, we may assume that $E \cap \Omega = \emptyset$ and hence $L_3 \cap E = L_4 \cap E = \emptyset$.

Case IB4a. D intersects L_2^-.

There is a fin F_2 in B_α for L_2 and L_4, whose edge is disjoint from E and whose base is a slit σ_{24} in $Bd\,B_\alpha$ from x_2 to x_4. Since $D \cap L_1 = \emptyset$ and $E \cap L_3 = \emptyset$, then by Lemma 6.4, we may assume $D \cap \delta_1 = \emptyset$ and $E \cap \delta_3 = \emptyset$. Thus we may also assume that $E \cap \sigma_{24} = \emptyset$. Then if L_2^- is disjoint from E, we have an S-case in B_α.

Suppose L_2^- intersects E. Then $L_2^+ \cap D = \emptyset$ and there is a fin F_1 in B_α for L_1 and L_2, whose edge is disjoint from D, and whose base τ_{12} is a slit in $Bd\,B_\alpha$ from x_1 to x_2 and such that $\tau_{12} \cap \sigma_{24} = \{x_2\}$. Since $E \cap L_4 = \emptyset$, we may also assume that $E \cap \delta_4 = \emptyset$. Thus we may assume that, in addition, $\tau_{12} \cap D = \emptyset$. We then have an S-case in B_α with $x_2 \in S$.

Case IB4b. E intersects L_2^- but D does not intersect L_2^-.

It is straightforward to show that we have an S-case in B_α.

Case IB4c. $L_2^- \cap (D \cup E) = \emptyset$.

First, we may assume that $C_\alpha \cap (D \cup E) = \emptyset$. Since $L_1^- \cap D = \emptyset$ and $L_3^- \cap E = L_4^- \cap E = \emptyset$, we have a situation similar to that of Case IB4a where both D and

E intersect L_2^-, but with B_α and C_α interchanged. Thus we have an S-case in C_α with $y_2 \in S$. Thus we may assume that $C_\alpha \cap (D \cup E) = \emptyset$.

Now we may assume that if μ is any curve of intersection of E with P, then $\mu \sim 0$ in $P - (\{x_2\} \cup D)$. This argument is the same as in Case IA. Thus by Lemma 5.1, we may assume that $E \cap B_\alpha = \emptyset$. This is Case 0.

This concludes Case IB since clearly there is no E-cap in X^3, and if K is any E-cap in C_α, then $Bd\,K$ does not separate y_3 and y_4 on $Bd\,C_\alpha$.

11. CASE II

Case II. $Bd\,H$ separates x_1, x_2, and x_3 from x_4 on $Bd\,B_\alpha$.

Case IIA. L_1^+, L_2^+, and L_3^+ intersect H.

Case IIA1. L_4^+ intersects H.

We may assume by Lemma 6.3 that $\delta_4^- \cap E = \emptyset$. Using H and δ_4, we may construct mutually disjoint polygonal simple closed curves J_1, J_2, and J_3 such that for each i, $i \neq 4$, $J_i \cap Bd\,B_\alpha = \{x_1, x_4\}$, and $J_i \subset L_i \cup H \cup \delta_4^- \cup L_4$. By Lemma 6.4, we may assume that $E \cap Bd\,B_\alpha = \emptyset$. This is Case 0.

Case IIA2. $L_4^- \cap H = \emptyset$.

By Lemma 6.1, there is an E-cap K in B_α or in C_α. Suppose K has base Q. If $K \subset B_\alpha$, then clearly $Q \subset \Omega$ and $x_4 \in Q$. If Y_3 is the 3-cell in B_α bounded by $K \cup Q$, we may use a half-fin in Y_3 to modify D and E. Thus we may assume that $L_4 \cap D = \emptyset$.

If $K \subset C_\alpha$, then $Bd\,K$ does not separate any of two of y_1, y_2, and y_3 on $Bd\,C_\alpha$, and we may take $y_4 \in Q$. By Lemma 6.2, L_4^- intersects K. By an argument like that of the preceding paragraph, we may assume $L_4 \cap D = \emptyset$.

By a construction similar to that used at the beginning of Case IB1, but using three fins, we may assume that $E \cap P = \emptyset$, and hence $L_1 \cap E = L_2 \cap E = L_3 \cap E = \emptyset$. We now have an S-case in B_α.

In all remaining cases, we shall assume that $L_1^+ \cap H = \emptyset$. If follows by Lemma 6.2 that $L_4^+ \cap H \neq \emptyset$. We may construct a half-fin F in X^3 for L_4 and H, modify D and E near F. Thus we may assume that $E \cap X^3 = \emptyset$ without changes in W^3 or C_α. Hence also $L_4 \cap E = \emptyset$.

Case IIB. L_2^+ and L_3^+ intersect H.

There is an E-cap K, with base Q, either in B_α or C_α.

Case IIB1. $K \subset B_\alpha$.

Clearly $q \in P$ and $x_1 \in Q$ but by Lemma 6.2, neither x_2 nor x_3 lies in Q. By using a half-fin in the 3-cell Y^3, we may assume $L_1 \cap D = \emptyset$.

First, we may assume that $C_\alpha \cap (D \cup E) = \emptyset$. To see this, we consider cases.

Suppose L_2^+ and L_3^+ are disjoint from E. We may construct a slit σ_{234} in $Bd\,B_\alpha$ containing x_2, x_3, and x_4 but not x_1, and may assume $E \cap \sigma_{234} = \emptyset$. We then have an S-case in B_α.

Suppose L_2^+ intersects E but $L_3^+ \cap E = \emptyset$. Then $L_2^- \cap (D \cup E) = \emptyset$. We may then construct a slit ω in $Bd\,B_\alpha$ from x_1 to x_4, containing x_2 and x_3 in the order $x_1 x_2 x_3 x_4$ on ω from x_1 to x_4, and may assume D is disjoint from the subarc $x_1 x_2$ of ω, and E is disjoint from the subarc $x_2 x_4$ of ω. We then have an S-case in B_α with $x_2 \in S$.

Suppose both L_2^+ and L_3^+ intersect E. Then L_2^- and L_3^- are disjoint from $D \cup E$, and we may construct an arc ω as above but on C_α from y_1 to y_4, containing y_2

and y_3, and assume that y_1y_2 is disjoint from D, y_2y_3 is disjoint from $D \cup E$, and y_3y_4 is disjoint from E. We then have an S-case in C_α where $y_2 \in S$ and $y_3 \in S$.

Hence we may assume that $C_\alpha \cap (D \cup E) = \emptyset$. We shall now show that if μ is any curve of intersection of E with P, then $\mu \sim 0$ in $P - (\{x_2, x_3\} \cup D)$. For if not, there is an E-cap K' for W^3 with base Q' in P such that $(Bd\,K') \nsim 0$ on $P - (\{x_2, x_3\} \cup D)$. Clearly $Q' \subset W^3$ and $x_1 \in Q'$ but neither x_2 nor X_3 belongs to Q'. It follows that there is a curve of intersection of D with $Bd\,B_\alpha$ lying in Q. By using a half-fin in the 3-cell y' in W^3 bounded by $K' \cup Q'$, we may assume that $Y' \cap D = \emptyset$. This argument may be repeated, and since D has only finitely many curves of intersection with $Bd\,B_\alpha$, this process terminates after finitely many steps. By Lemma 5.1, we may then assume that $E \cap B_\alpha = \emptyset$. This is Case 0.

Case IIB2. $K \subset C_\alpha$.

Suppose $y_1 \in Q$. Then by Lemma 6.2, neither y_2, y_3, nor y_4 lies in Q. By using a half-fin in C_α, we may assume that $L_1 \cap D = \emptyset$. This case is now essentially the same as Case IB1.

Case IIC. Neither L_1^+ nor L_2^+ intersects H, but L_3^+ intersects H.

Case IIC1. There is an E-cap K in B_α with base Q in P such that $x_1 \in Q$ and $x_2 \in Q$.

This is essentially Case I with D and E interchanged.

Case IIC2. There is an E-cap K in C_α with base Q such that $y_1 \in Q$ and $y_2 \in Q$.

Since $L_3 \cap E = L_4 \cap E = \emptyset$, then by a construction similar to one used in Case IB1, we may modify D and E in Y^3 only, and may assume that, in addition, $L_1 \cap D = L_2 \cap D = \emptyset$. We may now assume that we have an S-case in C_α. Here we have two cases, one where L_3^+ is disjoint from E, and the other where L_3^+ intersects E, somewhat as in Case IIB1. Hence we may assume that $C_\alpha \cap (D \cup E) = \emptyset$.

Now we may show, as in various previous cases, that if μ is a curve of intersection of E with P, then $\mu \sim 0$ in $P - (\{x_3\} \cup D)$. Then we may reduce this to Case 0.

Case IIC3. There exist (1) an E-cap K_1 with base Q_1 such that either (a) $K_1 \subset B_\alpha$, $x_1 \in Q_1$, but $x_2 \notin Q_1$, or (b) $K_1 \subset C_\alpha$, $y_1 \in Q_1$, but $y_2 \notin Q_1$, and (2) an E-cap K_2 with base Q_2 such that either (a) $K_2 \subset B_\alpha$, $x_2 \in Q_2$, but $x_1 \notin Q_2$, or (b) $K_2 \subset C_\alpha$, $y_2 \in Q_2$, but $y_1 \notin Q_1$.

The situations of this case are similar to those of the preceding two cases. This case is simpler than those, however, since we may use half-fins to get $D \cap L_1 = D \cap L_2 = \emptyset$.

Case IIC4. There is an E-cap K in B_α with base Q in P containing x_1 but not x_2, and (1) there exists no E-cap K_1 in B_α with base Q_1 in P containing x_2 and (2) there exists no E-cap K_2 in C_α with base Q_2 containing y_2.

This case is similar to Case IIB1, but there are more cases to consider. In all cases, we may assume that $L_1 \cap D = \emptyset$. If L_2^+ intersects one and only one of D and E, and L_3^+ intersects only D, then we may assume that we have S-cases in B_α. If L_2^+ intersects neither D nor E, and L_3^+ intersects only D, then we may assume that we have an S-case in B_α with $x_2 \in S$.

If L_2^+ intersects both D and E, and L_3^+ intersects only D, we may assume that we have an S-case in C_α with $y_2 \in S$. If both L_2^+ and L_3^+ intersect both D and E, we may assume that we have an S-case in C_α where both y_2 and y_3 are on S. In these two cases, then, we may assume that $C_\alpha \cap (D \cup E) = \emptyset$.

In the two cases of the paragraph above we may use the hypothesis of this case and the fact that L_3^+ intersects H to show that if μ is any curve of intersection of

E with P, then $\mu \sim 0$ in $P - (\{x_2, x_3\} \cup D)$.

The last subcase to be considered is that where L_3^+ intersects both D and E, and L_2^- intersects both D and E. We shall give an argument which also holds in Case IID3 below.

We shall show that if μ is any curve of intersection of E with $Bd\, B_\alpha$ lying in P, then $\mu \sim 0$ in $P - (\{x_2, x_3\} \cup D)$.

Suppose not. By the Loop Theorem [16], there is a polyhedral disc K' in $\mathbb{R}^3$ such that $Bd\, K' \subset P$, $\text{Int } K' \cap Bd\, B_\alpha = \emptyset$, and $Bd\, K' \nsim 0$ on $P - (\{x_2, x_3\} \cup D)$. Let Q' be the subdisc of P bounded by $Bd\, K'$. Then $x_4 \notin Q'$ but some x_i lies in Q'. Let Y' be the 3-cell in $\mathbb{R}^3$ bounded by $K' \cup Q'$.

First, suppose $K' \subset B_\alpha$. Then K' is an E-cap in B_α. Note that by Lemma 6.2, $x_2 \notin Q'$. If x_1 and x_3 lie in Q', we have a previous case. If only x_3 lies in Q', we may use a half-fin to modify D and E so that we have $L_3^+ \cap D = \emptyset$. This is a previous case. If only x_1 lies in Q', then Q' intersects D, and we may reduce the number of curves of intersection of D with $Bd\, B_\alpha$.

Now suppose $K' \not\subset B_\alpha$. Since $E \cap \Omega = \emptyset$, then $K' \cap B_\alpha = Bd\, K'$. Since $L_3^- \cap (D \cup E) = \emptyset$, then by the proof of Lemma 6.2, $x_3 \notin Q'$.

<u>Case IIC4a.</u> $x_2 \in Q'$. Since $L_3^- \cap Y' = \emptyset$, then L_2^- intersects K'. This case is similar to Case IB2, except that K' is not an E-cap. By an argument similar to the proof of Lemma 6.5, we may assume that $\delta_3^- \cap K' = \emptyset$.

Let ϵ_0 be the component of $(Bd\, \delta_2^-) - K'$ intersecting δ_3^-. Let ϵ_2 be a polygonal simple closed curve which is the union of ϵ_0 and an arc on K'. Then $\epsilon_2 \cap D = \emptyset$. We may assume that in addition $\delta_3^- \cap (D \cup E) = \emptyset$; the modifications made here are near δ_3^-. Thus there is a fin F_{23} in C_α for L_2 and L_3 with edge disjoint from D. To construct this fin, we use the fact that the point common to $Bd\, \delta_2^-$ and δ_3^- is not in Y', and that δ_3 does not intersect D.

It is easy to see that we have an S-case in C_α with $y_3 \in S$.

<u>Case IIC4b.</u> Only $x_1 \in Q'$.

It follows that D intersects Q', and by the Loop Theorem [16], there is a polyhedral disc H' such that $Bd\, H' \subset Q'$, $\text{Int } H'$ is disjoint from $K' \cup Q'$, and $Bd\, H' \nsim 0$ on $Q' - \{x_1\}$. Clearly H' cannot lie in Y'. Let P' be the disc in Q' bounded by $(Bd\, H')$; then $x_1 \in P'$. Let W' be the 3-cell in $\mathbb{R}^3$ bounded by $H' \cup P'$. Clearly $L_1^+ \subset W'$ and $L_1^- \cap W' = \emptyset$.

Since $L_2^+ \cap D = \emptyset$ and the loops of L_1^+ and L_2^+ are linked, it follows, as in the proof of Lemma 6.2, that $L_2^+ \subset W'$. If $x_3 \notin W'$, then we may give an argument similar to that for Case IIC4a above, but with B_α and C_α interchanged, to show that we have an S-case in B_α. Thus $x_3 \in W'$ and since $L_3^- \cap D = \emptyset$, then $L_3^- \subset W'$. But since $L_1^- \cap W' = \emptyset$ and the lower loops of L_1^- and L_3^- are linked, this is a contradiction.

<u>Case IIC5.</u> As Case IIC4 above but with $K \subset C_\alpha$.

We may assume that $L_1 \cap D = \emptyset$. Then this case is similar to Case IIC4.

<u>Case IID.</u> No one of L_1^+, L_2^+, and L_3^+ intersects H.

<u>Case IID1.</u> If $i = 1, 2$, or 3, there is an E-cap K_i, in B_α or C_α, with base Q_i containing x_i.

We may assume that if there is more than one K_i, then any two are disjoint.

Suppose some Q_i, either on $Bd\, B_\alpha$ or $Bd\, C_\alpha$, contains either two or three x's. Then by an argument similar to that for Case IB1, we may assume that each of the corresponding L's is disjoint from D. If Q_i contains only x_i among the x's, we may use a half-fin to modify D and E. Thus we may assume in this case that if

$i = 1, 2,$ or 3, $L_i \cap D = \emptyset$.

By constructing fins in B_α for L_1 and L_2, and L_2 and L_3, we may assume that there is a slit τ_{123} in $Bd\,B_\alpha$ containing x_1, x_2, x_3, but not x_4, and that $D \cap \tau_{123} = \emptyset$. We now have an S-case in B_α.

<u>Case IID2.</u> If $i = 1$ or 2, there exists an E-cap K_i with base Q_i, either in $Bd\,Q_\alpha$ containing x_i or in $Bd\,C_\alpha$ containing y_i, but no E-cap K with base Q, either in B_α containing X_3 or in C_α containing y_3.

As in previous cases, we may assume that $L_1 \cap D = L_2 \cap D = \emptyset$.

<u>Case IID2a.</u> L_3^- intersects D but not E.

By using a fin in B_α for L_1 and L_2, we may assume that there is a slit τ_{12} in $Bd\,B_\alpha$ from x_1 to x_2 and disjoint from D. By using a fin in B_α for L_3 and L_4, we may assume that there is a slit σ_{34} in $Bd\,B_\alpha$, from x_3 to x_4 and disjoint from τ_{12}, such that $\sigma_{34} \cap E = \emptyset$. Then we have an S-case in B_α.

<u>Case IID2b.</u> L_3^- intersects both D and E.

Then $L_3^+ \cap (D \cup E) = \emptyset$. We make a construction as in Case IID2a above, but since $L_3^+ \cap (D \cup E) = \emptyset$, we may assume there is a slit ω_{23} from x_2 to x_3 in $Bd\,B_\alpha$ such that $\omega_{23} \cap (D \cup E) = \emptyset$ and $(\text{Int } \omega_{23}) \cap (\sigma_{34} \cup \tau_{12}) = \emptyset$. We then have an S-case in B_α with $x_3 \in S$.

<u>Case IID2c.</u> L_3^- intersects E but not D.

This is similar to Case IID1.

<u>Case IID2d.</u> L_3^- intersects neither D nor E.

First, we may assume that $C_\alpha \cap (D \cup E) = \emptyset$. Since $L_1 \cap D = L_2 \cap D = \emptyset$, $L_3^- \cap (D \cup E) = \emptyset$, and $L_4 \cap E = \emptyset$, this is like Case IID2b but with B_α replaced by C_α. Then we may assume we have an S-case in C_α with $y_3 \in S$.

Now we shall prove that if μ is any curve of intersection of E with P, then $\mu \sim 0$ in $P - (\{x_3\} \cup D)$. If not, there exists an E-cap K'-cap K' in B_α with base Q' in P such that $Bd\,K' \nsim 0$ in $P - (\{x_3\} \cup D)$. By the hypothesis of this case, $x_3 \notin Q$. Hence there is a curve of intersection of D with $Bd\,B_\alpha$ in Q'. By an argument like that for Case IB1, we may reduce the number of such curves. We may therefore reduce this case to Case 0.

<u>Case IID3.</u> There exists an E-cap K with base Q either in $Bd\,B_\alpha$ containing x_1 or in $Bd\,C_\alpha$ containing y_1, but if $i = 2$ or 3, there is no E-cap K' with base Q' either in $Bd\,B_\alpha$ containing x_i or in $Bd\,C_\alpha$ containing y_i.

This case is similar to Case IIC4 and Case IIC5. Here, however, there is an additional subcase to be considered. This is the case where L_2^+ intersects niether D nor E, and L_3^+ intersects neither D nor E. In this case, we may assume that we have an S-case in B_α with both x_2 and x_3 on S.

The various other cases that may occur when there is a D-cap in B_α may be handled by arguments similar to those above. If there is an E-cap in B_α, we interchange D and E. □

Lemma 11.1. *Suppose the hypothesis of Lemma 9.1. Then there exist admissible modifications D'' and E'' of D and E, respectively, with respect to T_α such that*

(1) $D'' \cap E'' \cap T_\alpha = \emptyset$,

(2) $(D'' \cap E'') \cap (B_\alpha \cup C_\alpha) = \emptyset$, *and*

(3) *if $i = 1, 2, 3,$ or 4, L_i has Property $\overline{P}$ with respect to D'' and E''.*

Proof. Let U and V be disjoint neighborhoods of B_α and C_α, respectively, with $U \cup V \subset T_\alpha$. By Lemma 9.1, there exist modifications D' and E' of D and E, respectively, with properties as stated in the conclusion of Lemma 9.1. By a lemma like Lemma 9.1 but for C_α and V, there exist modifications D'' and E'' of D' and E', respectively, with properties analogous to those above. Since $U \cap V = \emptyset$, then $(D'' \cup E'') \cap C_\alpha = \emptyset$, and $D'' \subset D' \cup V$ and $E'' \subset E' \cup V$. Since $B_\alpha \subset U, U \cap V = \emptyset$, and $(D' \cup E') \cap B_\alpha = \emptyset$, it follows that $D'' \cap B_\alpha = E'' \cap B_\alpha = \emptyset$. $\square$

12. Graphs and Cores

There are two main steps in the proof of the Generalized Bing Lemma. The first of these is provided by Lemma 11.1. We turn now to the second, which will be provided by Lemma 12.2. We shall then prove the Generalized Bing Lemma.

Suppose α is an index. A polygonal copy L'_α in T_α of the core L_α of T_α is *parallel* to L_α if and only if there is a piecewise linear embedding $f : L_\alpha \times [0,1]$ into T_α such that $f(L_\alpha \times \{0\}) = L'_\alpha$ and $f(L_\alpha \times \{1\}) = L_\alpha$.

Lemma 12.1. *Suppose α is an index, and D and E are polyhedral singular discs in $\mathbb{R}^3$ such that*

(1) $(Bd\,D) \cup (Bd\,E)$ *and T_α are disjoint, and*
(2) $D \cap E \cap T_\alpha = \emptyset$.

Suppose L'_α is a polygonal copy of L_α in T_α parallel to L_α such that the stem of L'_α is disjoint from $D \cup E$. Suppose $f : L_\alpha \times [0,1] \to T_\alpha$ is a piecewise linear embedding of $L_\alpha \times [0,1]$ into T_α such that $f(L_\alpha \times \{0\}) = L'_\alpha$ and $f(L_\alpha \times \{1\}) = L_\alpha$. Let $f(L_\alpha \times [0,1]) = K$. Then there exist admissible modifications D' of D and E' of E, respectively, such that

(1) $D' \cap E' \cap T_\alpha = \emptyset$
(2) $D' \cap L'_\alpha \subset D \cap L'_\alpha$ *and* $E' \cap L'_\alpha \subset E \cap L'_\alpha$, *and*
(3) *if M is any component of $(D' \cup E') \cap K$, then for some point x of L'_α, $M = f(\{x\} \times [0,1])$.*

Proof. Let σ be the stem of L'_α. Then since D and E are disjoint form the stem of L'_α, we may assume that D and E are disjoint from the disc $f(\sigma \times [0,1])$. This reduces the problem to that of modifying the intersections of D and E with certain discs. For this purpose, we may use disc-swapping and the creasing operation of the proof of Lemma 7.2. $\square$

Suppose that α is an index. Let p_α be the center point of B_α, and if $i = 1, 2, 3$, or 4, let $A'_{\alpha i}$ be the interval in B_α from p_α to $x_{\alpha i}$. Let q_α be the center point of C_α, and if $i = 1, 2, 3$, or 4, let $A''_{\alpha i}$ be the interval in C_α from q_α to $y_{\alpha i}$. If $i = 1, 2, 3$, or 4, let $A_{\alpha i}$ be the arc which is the union of $A'_{\alpha i}$, $A''_{\alpha i}$, and the arc of the stem of $L_{\alpha i}$ from $x_{\alpha i}$ to $y_{\alpha i}$. See Figure 5. Note that the graph $\cup_{i=1}^4 A_{\alpha i}$ lies slightly to one side of the plane Π_α.

The following lemma is an adaptation to our situation of Part II of the proof of Theorem 10 of [8].

Lemma 12.2.. *Suppose α is an index, D and E are polyhedral singular discs in $\mathbb{R}^3$ such that*

(1) $(Bd\,D) \cup (Bd\,E)$ *and T_α are disjoint and*
(2) $D \cap E \cap T_\alpha = \emptyset$. *Suppose that each of $A_{\alpha 1}, A_{\alpha 2}, A_{\alpha 3}$, and $A_{\alpha 4}$ intersects at most one of D and E.*

Then there exist admissible modifications D' of D and E' of E', respectively, relative to T_α, such that

(1) $D' \cap E' \cap T_\alpha = \emptyset$ *and*
(2) *the core L_α of T_α has Property $\overline{P}$ with respect to D' and E'.*

Proof. There are several cases. For each i, let A_i denote $A_{\alpha i}$.

Case 1. A_1, A_2, and A_3 are disjoint from E. First we shrink A_1 to a point. Near the resulting graph, there is a polygonal copy L'_α of L_α, parallel to L_α, disjoint from E, and with its stem disjoint from D and E.

Case 2. A_1 and A_2 are disjoint from D, and A_3 and A_4 are disjoint from E. By a *pinched cylinder* we shall mean a space obtained from $S^1 \times [0,1]$ by collapsing $\{t\} \times [0,1]$ for some point t on S^1. We may adjust the A's slightly near p_α so that there is a polyhedral pinched cylinder C on T_α whose boundary is the pair of simple closed curves $A_1 \cup A_2$ and $A_3 \cup A_4$; note that the two curves are wedged at q_α. It follows by theorems of plane topology that there is a polygonal simple closed curve J in C, containing q_α, disjoint form $D \cup E$, and homotopic in C, with q_α fixed, to $A_1 \cap A_2$. We may now use $J \cup A_2 \cup A_4$ to construct a polygonal copy L'_α of L_α in T_α, parallel to L_α, such that (1) the stem and the lower loop of L'_α are disjoint from both D and E, and (3) one side of the upper loop of L'_α is disjoint from D and the other side is disjoint from E.

Case 3. A_1 and A_4 are disjoint from D, and A_2 and A_3 are disjoint from E. We may adjust the $A's$ slightly near q_α so that there is a polyhedral pinched cylinder C in T_α whose boundary is the pair of simple closed curves $A_1 \cup A_3$ and $A_2 \cup A_4$; note that these curves are wedged at p_α. It follows by theorems of plane topology that there is a polygonal simple closed curve J in C, containing p_α, homotopic in C with p_α fixed to $A_1 \cup A_3$, and disjoint from one of D and E. If J is disjoint from D, we may use $A_1 \cup A_4 \cup J$ to construct a polygonal copy L'_α of L_α in T_α parallel to L_α and disjoint from D. If J is disjoint from E, we may use $A_2 \cup A_3 \cup J$ to construct a polygonal copy L'_α of L_α in T_α parallel to L_α and disjoint from E. In each case we may assume that the stem of L_α is also disjoint from D.

Other cases are similar to those above. In all cases, by Lemma 12.1, there exist admissible modifications D' and E' of D and E, respectively, relative to T_α, such that (1) $D' \cap E' \cap T_\alpha = \emptyset$ and (2) L_α has Property $\overline{P}$ with respect to D' and E'. $\square$

Lemma 12.3. *Suppose that α is any index, and D and E polyhedral singular discs in $\mathbb{R}^3$ such that*

(1) $(Bd\,D) \cup (Bd\,E)$ *and T_α are disjoint, and*
(2) $D \cap E \cap T_\alpha = \emptyset$.

Suppose that if D' and E' are any admissible modifications of D and E, respectively, with respect to T_α, such that $D' \cap E' \cap T_\alpha = \emptyset$, then L_α has Property P with respect to D' and E'. Then there exist an index i such that

(1) $i = 1, 2, 3,$ *or 4 and*
(2) *if D'' and E'' are any admissible modifications of D and E, respectively, relative to $T_{\alpha i}$ such that $D'' \cap E'' \cap T_{\alpha i} = \emptyset$, then $L_{\alpha i}$ has Property P with respect to D'' and E''.*

Proof. Suppose that there exist an index α and polyhedral singular discs D and E satisfying the hypothesis, but no index i satisfying the conclusion. By Lemma 4.1, there exist admissible modifications $\widehat{D}$ and $\widehat{E}$ of D and E, respectively, relative to

T_α, such that (1) $\widehat{D} \cap \widehat{E} \cap T_\alpha = \emptyset$ and (2) $i = 1, 2, 3,$ or 4, $L_{\alpha i}$ has Property $\overline{P}$ with respect to $\widehat{D}$ and $\widehat{E}$.

By Lemma 11.1, there exist admissible modifications D' and E' of $\widehat{D}$ and $\widehat{E}$, respectively, relative to T_α, such that (1) $D' \cap E' \cap T_\alpha = \emptyset$, (2) neither D' nor E' intersects either B_α or C_α, and (3) if $i = 1, 2, 3,$ or 4, $L_{\alpha i}$ has Property $\overline{P}$ with respect to D' and E'. It follows that if $i = 1, 2, 3,$ or 4, $L_{\alpha i}$ intersects at most one of D' and E', and hence that the arc $A_{\alpha i}$ intersects at most one of D' and E'.

By Lemma 12.2, there exist admissible modifications D'' and E'' of D' and E', respectively, relative to T_α, such that (1) $D'' \cap E'' \cap T_\alpha = \emptyset$ and (2) the core L_α of T_α has Property $\overline{P}$ with respect to D'' and E''. By Lemma 4.1, D'' and E'' are admissible modifications of D and E, respectively, relative to T_α. This is a contradiction. $\square$

Lemma 12.4. (Generalized Bing Lemma) *Suppose α is an index, D and E are polyhedral singular discs in $\mathbb{R}^3$ such that $Bd\,D$ links the upper handle of T_α, $Bd\,E$ links the lower handle of T_α, and $D \cap E \cap T_\alpha = \emptyset$. Then some element of G in T_α intersects both D and E.*

Proof. Suppose D' and E' are any admissible modifications of D and E, respectively, relative to T_α, such that $D' \cap E' \cap T_\alpha = \emptyset$. Since $Bd\,D = Bd\,D'$ and $Bd\,E = Bd\,E'$, then $Bd\,D'$ links the the upper handle of T_α and hence D' intersects the upper loop of L_α, and $Bd\,E'$ links the lower handle of T_α and hence E' intersects the lower loop of L_α. Thus L_α has Property P with respect to D' and E'. By Lemma 12.3, there is an index i_1 such that (1) $i_1 = 1, 2, 3,$ or 4 and (2) if D'' and E'' are any admissible modifications of D and E, respectively, relative to $T_{\alpha i_1}$ such that $D'' \cap E'' \cap T_{\alpha i_1} = \emptyset$, then $L_{\alpha i_1}$ has Property P with respect to D'' and E''.

Suppose that n is some positive integer and $i_1, i_2, \cdots,$ and i_n have been chosen. Then by an argument like that above, using Lemma 12.3, we may show that there exists an index i_{n+1} such that (1) $i_n = 1, 2, 3,$ or 4, and (2) if D'' and E'' are any admissible modifications of D and E, respectively, relative to $T_{\alpha i_1 i_2 \cdots i_n i_{n+1}}$, then $L_{\alpha i_1 i_2 \cdots i_n i_{n+1}}$ has Property P with respect to D'' and E''. It follows that for each n, i_n is defined.

Let n be any positive integer, and let $\beta = \alpha i_1 i_2 \cdots i_n$. Clearly D and E are admissible modifications of D and E, respectively, relative to T_β. Thus L_β has Property P with respect to D and E. Thus L_β intersects both D and E, and hence, so does T_β.

Let $g = \cap_{n=1}^\infty T_{\alpha i_1 i_2 \cdots i_n}$. Then g is an element of G in T_α, and clearly g intersects both D and E. $\square$

13. THE MAIN RESULT

The main result of this paper may be established by an argument that is closely related to the proof of the main result of [5]. A subset M of $\mathbb{R}^3$ is *saturated* if and only if each element of G intersecting M lies in M. If α is an index, then Λ_α is a *pseudo core* of T_α if and only if Λ_α is a polygonal figure eight in $Int\,T_\alpha$ such that one loop (the *upper loop*) of Λ_α is nontrivial in $T_\alpha \cup C_\alpha$, and the other loop (the *lower loop*) of Λ_α is nontrivial in $T_\alpha \cup B_\alpha$.

The following lemma provides the main step in the proof of the theorem of this section.

Lemma 13.1. *Suppose α is an index and $i = 1, 2, 3,$ or 4. Suppose U is a simply connected saturated open set in $\mathbb{R}^3$ such that U contains a pseudocore of $T_{\alpha i}$. Then U contains a pseudocore of T_α.*

Proof. In [5], the proof of the corresponding result, Lemma 2, is broken into two cases. The first, <u>Case</u> A, is treated in Lemma 3 of [5]. We may define <u>Case</u> A in this paper in an analogous way, and the proof of Lemma 3 of [5] applies.

If <u>Case</u> A fails, then in [5], Lemma 5 is used to establish Lemma 2. We may establish, for our situation here, an analogue to Lemma 5. In place of the argument of pages 47, 48, and the top half of page 49 of [5], we apply Lemma 12.4 above. Then the proof of Lemma 2 of [5] gives a proof of Lemma 13.1. □

Let $\varphi : \mathbb{R}^3 \longrightarrow \mathbb{R}^3/G$ be the projection map.

Theorem 13.2. *If G is Bing's dogbone decomposition of $\mathbb{R}^3$, then $\mathbb{R}^3/G$ is not strongly locally simply connected. In fact, if g is any nondegenerate element of G, there is no simply connected open set in $\mathbb{R}^3/G$ containing $\varphi(g)$ and lying in $\varphi(Int\ T_0)$.*

Proof. The proof of the Theorem of [5] may be adapted to give a proof of Theorem 13.2. □

References

1. S. Armentrout, *On the strong local simple connectivity of the decomposition spaces of toroidal decompositions,* Fund. Math. **69** (1970), 15–37.
2. _____, *A decomposition of E^3 into straight arcs and singletons,* No. 73 (1970), 49 pp, Dissertationes Mathematicae.
3. _____, *A three-dimensional spheroidal space which is not a sphere,* Fund. Math. **68** (1970), 183–186.
4. _____, *On the singularity of Mazurkiewicz in absolute neighborhood retracts,* Fund. Math. **69** (1970), 131–145.
5. _____, *Local properties of knotted dogbone spaces* **24** (1986), 41–52.
6. _____, *Cellular extensions of cellular decompositions,* to appear.
7. R. H. Bing, *Decomposition of E^3 into points and tame arcs,* In "Summary of Lectures and Seminars," Summer Institute on Set Theoretic Topology, Madison (1955), Revised (1958), 41–48, University of Wisonsin.
8. _____, *A decomposition of E^3 into points and tame arcs such that the decomposition space is topologically distinct from E^3,* Ann. of Math. **65** (1957), 484–500.
9. _____, *Point-like decompositions of E^3,* Fund. Math. **50** (1962), 431–453.
10. _____, *Decompositions of E^3,* Topology of 3-Manifolds and Related Topics (M. K. Fort, Jr., editor), 5–21, Prentice-Hall, Englewood Cliffs, N. J., 1962.
11. M. L. Curtis, Topology of 3-Manifolds and Related Topics (M. K. Fort, JR., editor) Prentice-Hall, Englewood Cliffs, N. J., 1962.
12. M. K. Fort, Jr., *A note concerning a decomposition space defined by Bing,* Ann. of Math. **65** (1957), 501–504.
13. E. Hutchings, Thesis, Univ. of British Columbia, 1968.
14. H. W. Lambert, *A topological property of Bing's decomposition of E^3 into points and tame arcs,* Duke Math J. **34** (1967), 501–510.
15. C. D. Papakyriakopoulos, *On solid tori,* Proc. London Math. Soc. **7(3)** (1957), 281–299.
16. J. R. Stallings, *On the loop theorem,* Ann. of Math. **72** (1960), 12–19.

Department of Mathematics, Penn State University, University Park, PA 16802
E-mail address: sxa@math.psu.edu

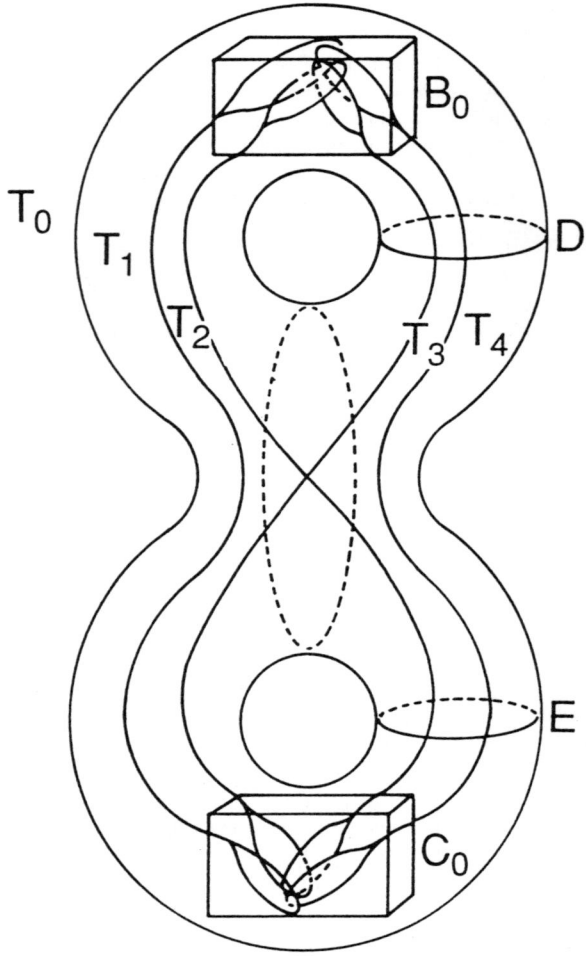

Figure 1

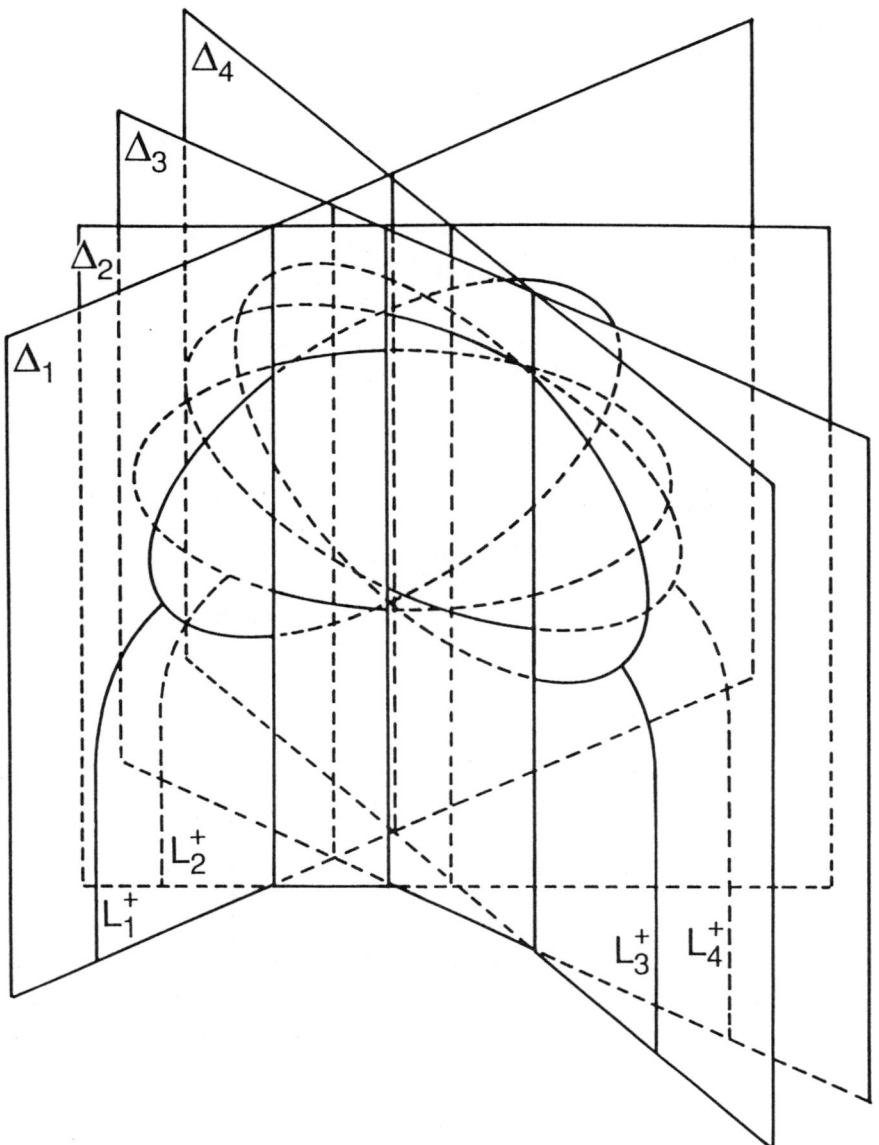

Figure 2

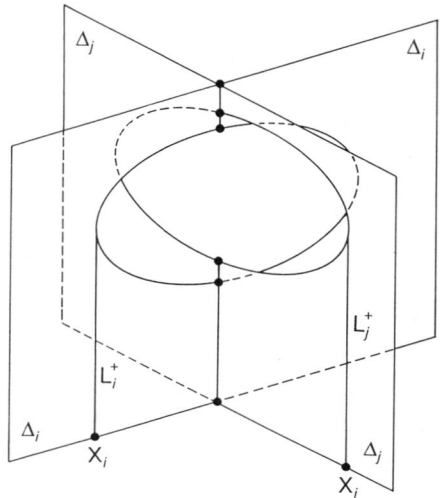

Figure 3 (a)

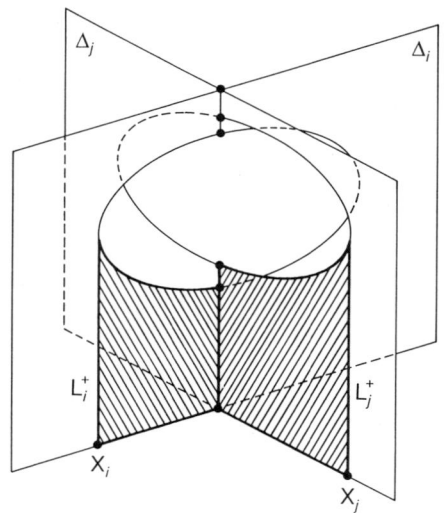

Figure 3 (b)

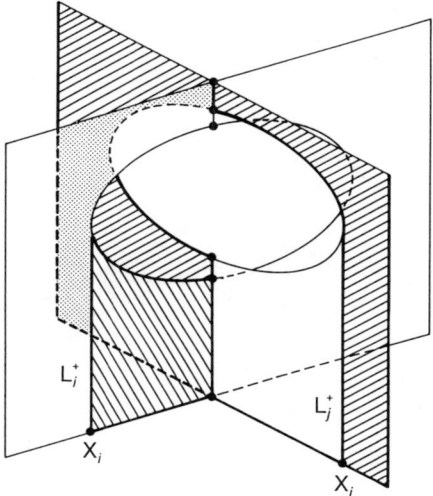

Figure 3 (c)

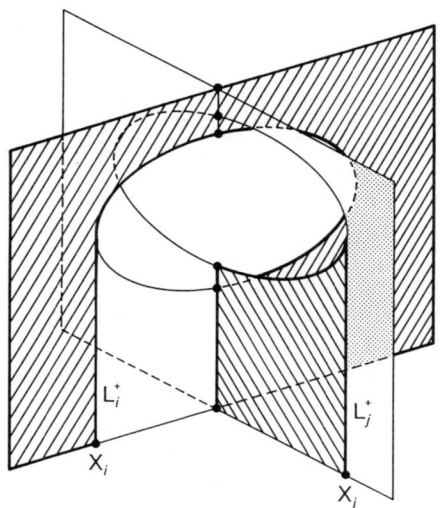

Figure 3 (d)

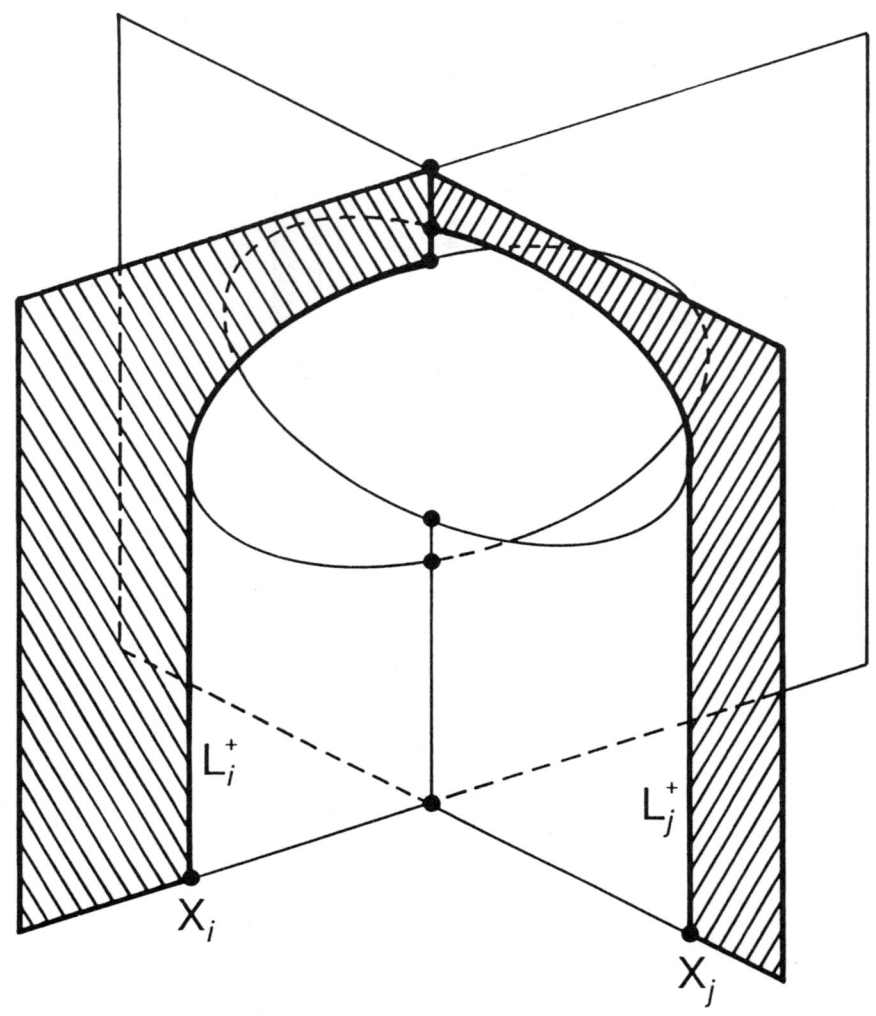

Figure 3 (e)

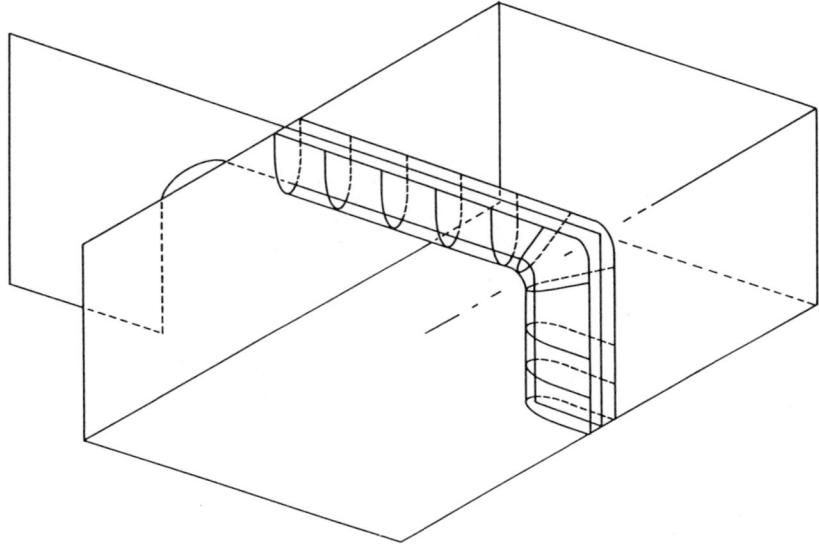

Figure 4 (a)

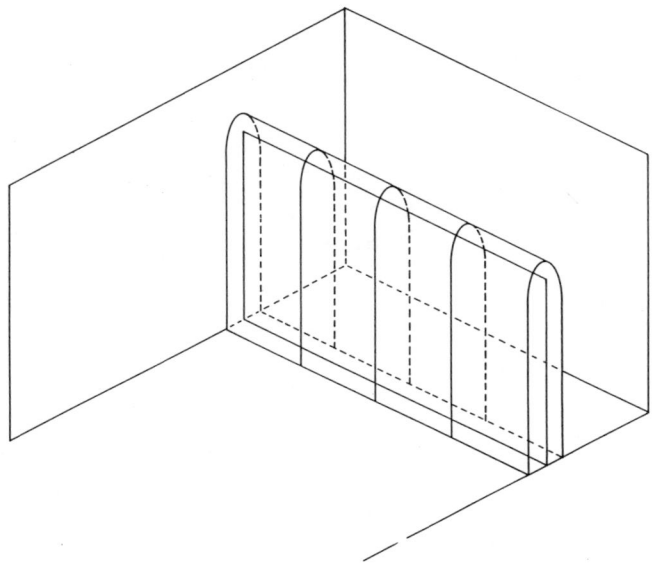

Figure 4 (b)

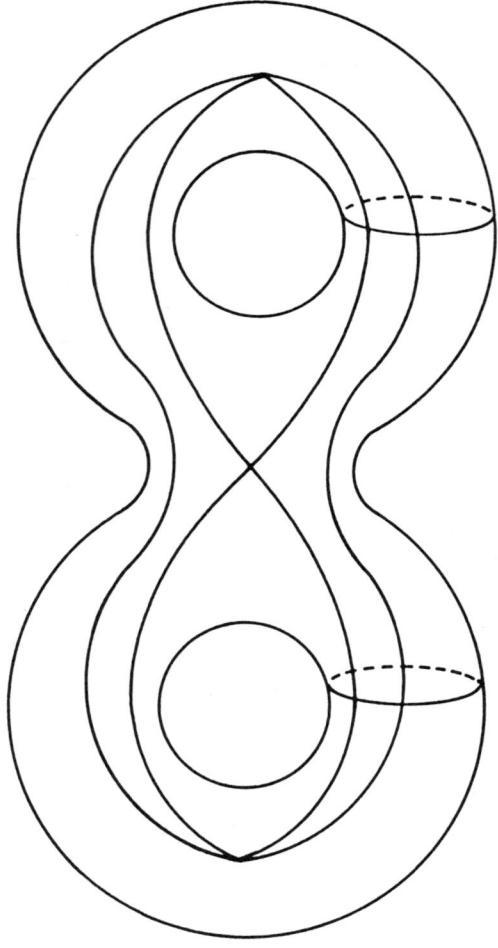

Figure 5

A PROGRAM FOR THE POINCARÉ CONJECTURE
AND SOME OF ITS RAMIFICATIONS

V. POÉNARU

1. INTRODUCTION

The text which follows is an expanded version of the lecture I gave at the Penn State Conference on low-dimensional topology in the honour of Steve Armentrout, in May 1996. Very roughly speaking my program for the Poincaré Conjecture consists of three steps:

Step 1. For any homotopy 3-ball Δ^3 we can find a compact smooth 4-manifold X^4 which has no handles of index one and which can be interpolated between $\Delta^3 \times I$ and a larger copy (of $\Delta^3 \times I$) obtained by adding a collar to the boundary

$$\Delta^3 \times I \subset X^4 \subset (\Delta^3 \times I) \bigcup_{\bullet} (\partial(\Delta^3 \times I) \times [0,1]). \qquad (1.1)$$

Step 2. (Which, for the time being, is still only conjectured). From the conclusion of Step 1, deduce that $\Delta^3 \times I$ itself has no handles of index one, in the smooth category.

Step 3. Let Δ^3 be a homotopy 3-ball for which it is assumed that $\Delta^3 \times I$ has no handles of index one, in the smooth category. Then $\Delta^3 = B^3$. (This also means that the "stable Poincaré Conjecture" implies the Poincaré Conjecture.)

Step 1, the complete proof of which is provided by [Po1], [Po2], [Po3], [Po4] will be briefly discussed in section 2 of this paper, where it appears as Corollary 3. For Step 2, the complete proof is provided by [Po5]; actually only the first two parts of this paper are available as preprints, the rest is in the process of typing, or just handwritten and waiting to be typed. This step is discussed in section 3 of this paper, where it appears as Corollary 6.

The whole program is discussed, in more detail than in the present lecture, in David Gabai's paper [Ga] (see also [Po0].) I think David's paper is the best introduction to the subject.

The ideas used in the program for the Poincaré Conjecture (or, at least, some of these ideas) have turned out to be quite useful in a different area, namely for the question of simple-connectivity at infinity for open simply-connected 3-manifolds V^3, the most interesting case being, of course, $V^3 = \widetilde{M}^3$, the universal covering space of a closed 3-manifold M^3 with infinite π_1. The papers [Po6], [Po7], [Po8], [Po9] are a good introduction to this part of the story.

2. THE SMOOTH TAMENESS THEOREM

The theorem which I will explain next, applies to any homotopy 3-sphere Σ^3. The associated homotopy-ball Σ^3-int B^3, will be denoted by Δ^3.

Theorem 1. (The smooth tameness theorem) *"For any homotopy 3-sphere Σ^3 one can find a smooth non-compact 4-manifold V^4 with non-empty boundary, such that*

$$\text{int } V^4 = R^4_{\text{standard}} \quad , \quad \partial V^4 = \sum_1^\infty S^1_i \times \text{int } D^2_i, \tag{2.1}$$

and with the following two properties:

(A)   *We have a diffeomorphism of open 4-manifolds*

$$\text{int } ((\Delta^3 \times I) \# \infty \# (S^2 \times D^2)) = V^4 + \{ \text{ the infinitely}$$
$$\text{many 2-handles } \sum_1^\infty D^2_i \times \text{int } D^2_i \text{ corresponding to } \partial V^4 \}. \tag{2.2}$$

[*The 2-handles in quesiton are without lateral boundary, "#" means connected sum along the boundary, and the parmeterization of ∂V^4 (2.1) corresponds to the null-framing.*]

(B)   *For any finite $N \in Z_+$, the truncation*

$$V^4 | N \underset{\text{def}}{=} V^4 - \sum_{N+1}^\infty S^1_i \times \text{int } D^2_i \tag{2.3}$$

can be smoothly compactified into a copy of B^4_{standard}."

Here is a list of *comments* and *explanations* concerning the statement above.

(1)   We will call *sort of link* a smooth non-compact manifold W^4 such that, for some $1 \le \alpha \le \infty$

$$\text{int } W^4 = R^4_{\text{standard}} \quad , \quad \partial W^4 = \sum_1^\alpha S^1_i \times +i^2 \text{ (null-framing)}. \tag{2.4}$$

So, our V^4 from the theorem is a sort of link, with $\alpha = \infty$.

Since we work in the **DIFF** category (which, as it will turn out, is essential for our approach) we have to specify in (2.1), (2.4), that the R^4 is standard; remember that, as a consequence of the work of M. Freedman [Fr1] and S. Donaldson [D], exotic R^4's do indeed exist [Fr2]. For the smooth B^4 it is unknown, at the present time, if exotic versions exist.

Here is the most obvious example of a sort of link (and, at the same time, the justification for the name). Consider, to begin with, a smooth pair

$$\left(B^4, \sum_1^{N<\infty} S^1_i \times D^2_i \subset \partial B^4 \right) \tag{2.5}$$

(with B^4, from now on, the standard smooth 4-ball). With this, the manifold

$$W^4 = B^4 - \left(\partial B^4 - \sum_1^N S^1_i \times \text{int } D^2_i \right) \tag{2.6}$$

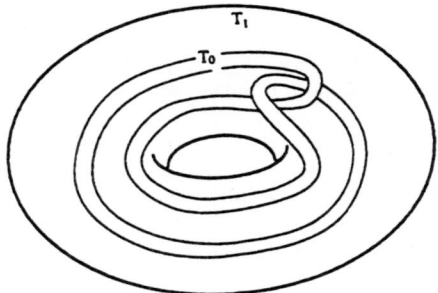

FIG. 1. THE WHITEHEAD PAIR $T_0 \subset$ int T_1

is a sort of link. Moreover, a sort of link of this particular type will be said to be *smoothly tame*. According to B) in our theorem, each $V^4 | N$ is smoothly tame.

(2) Here is an example of a sort of link which is *smoothly wild*. Start with the Whitehead pair of solid tori $T_0 \subset T_1$ from figure 1, and then iterate indefinitely the same embedding:

$$T_0 \subset T_1 \subset T_2 \subset T_3 \subset \ldots \ldots$$

This leads to the classical Whitehead manifold [Wh1]

$$W h^3 = \overset{\infty}{\underset{0}{\cup}} \, T_n \, . \tag{2.7}$$

It is a standard fact that for this open contractible manifold, we have the diffeomorphism

$$W h^3 \times (0,1) = \overset{\infty}{\underset{0}{\cup}} \, T_n \times (0,1) = R^4_{\text{standard}} \tag{2.8}$$

With a slight modification of this formula, we can get a sort of link, namely

$$W^4 \underset{\text{def}}{=} \{(\text{int } T_0) \times (0,1]\} \cup \overset{\infty}{\underset{1}{\bigcup}} T_n \times (0,1), \tag{2.9}$$

with $\partial W^4 = (\text{int } T_0) \times 1$. It turns out that this "sort of knot" is a Casson Handle [Ca1], [Fr1], [F-Q] which is known to be smoothly wild [B]. Of course, it is known from Freedman's work [Fr1] that all Casson Handles are topologically standard, and hence topologically tame. Via Donaldson's work [D] Casson Handles which are not smoothly standard (and hence not smoothly tame either) have to exist [G]. Some explicit ones are known already to be smoothly wild and presumably all are; I suspect that appropriate topological quantum field thories [Q] could be useful for this problem. [P-T4]. (see also [Fu1]). All Casson Handles admit description of type (2.9).

(3) The proof of the smooth tameness theorem is given in [Po4] and this paper also uses [Po1], [Po2], [Po3]. I will not even try to sketch the proof in question, for the time being, but only offer some comments. The argument makes heavy use of an *infinite process*, which is quite different from the one used by M. Freedman for the topological 4-dimensional Poincaré Conjecture [Fr1], [F-Q]. For instance, since we want to stay inside the smooth category, no Bing type shrinking can every be used in our context.

(4) Here is the reason why we want to get smooth results, rather than just topological ones. In the proof of the stange compactification theorem (see the next section of this paper) one starts with two distinct triangulations τ_1, τ_2 for $\Delta^3 \times I$, each with its own particular virtues: τ_1 is compatible with the product sructure, while the 2-skeleton of τ_2 is a collapsible 2-complex with 2-cells added. We need a triangulation τ with both properties and this would be possible *if* τ_1 and τ_2 would have a common subdivision. But we are in dimension four where the (general, **TOP**) Hauptvermutung is glamorously false. C. Taubes shows, for instance, that R^4 has uncountably many distinct **PL** structures [Ta]; this is very different indeed, from the high-dimensional situation [Su], [K-S], [Ca2]. But if both τ_1 and τ_2 are compatible with the **DIFF** structure of $\Delta^3 \times I$, then we can safely use Whitehead's smooth Hauptvermutung [Wh2], which is true without any dimensional restrictions.

(5) The infinite process via which Theorem 1 can be proved, constructs V^4 by putting together infinitely many compact non-simply-connected pieces, a bit like in (2.7). Also, for each individual finite truncation $V^4|N$, the same infinite process adds infinitely many pieces at infinity, until the boundary extends to a copy of R^3. It turns out that this is enough for compactifying $V^4|N$ into a copy of B^4. The reason is the following easy *fact*: In the smooth category, for $n \geq 4$ (in particular for $n = 4$ which is the case of interest for us), there is a unique way to glue R^{n-1} to the infinity of R^n

$$R^n \cup R^{n-1} = R^n_+;$$

(all the euclidean spaces considered here are standard.)

Notice, on the other hand, that for $n = 3$ the analogous fact is false, due to the existence of the Artin-Fox arcs. This will turn out to be a major difficulty, to be overcome, for the proof of the strange compactification theorem (see section 3, below.)

(6) It is not claimed that the smooth compactifications of $V^4|N$, $V^4|(N + 1)$ are compatible, and hence not that V^4 itself is smoothly tame either. It is hoped, on the other hand, that a more refined version of the present proof [Po4] will provide the compatibility relations

{Smooth compactification of $V^4|(N + 1)$, restricted to $V^4|N$} =

= Smooth compactification of $V^4|N$,

for all N's. such equalities, for all N, would be very useful in handling the conjectured Step 2 of our program, i.e. for closing the present gap betwen Steps 1 and 3 (see also the end of this section.) □

I will give now the main applications of the smooth tameness theorem. Some definitions will have to be given first. A smooth n-manifold X^n will be said to be *geometrically simply connected* if it possesses a smooth **PROPER** handlebody decomposition *without handles of index* $\lambda = 1$. [For the present purposes, X^n is compact and so the word "**PROPER**" is unnecessary. We will come back to the general case in section 4.] Let Y^n be a smooth n-manifold (which is not closed). We will say that Y^n is geometrically simply-connected *at long distance*, if for every compact subset $K \subset Y^n$ we can find a compact geometrically simply-connected submanifold $X^n \subset Y^n$, such that $K \subset X^n \subset Y^n$. In the special case when Y^n is itself compact bounded, there is an equivlaent form of this definition, where only compact objects occur. Let $\partial Y^n \times [0,1] \subset Y^n$ be a collar of the boundary $\partial Y^n = \partial Y^n \times 1$; define

$$Y^n_{\text{small}} \subset X^n - \partial Y^n \times (0,1], \tag{2.10}$$

i.e. Y^n_{small} is another version of Y^n, canonically embedded inside int Y^n. Then, Y^n is geometrically simply-connected at long distance iff there exists a compact geometrically simply-connected submanifold $X^n \subset Y^n$, such that

$$Y^n_{\text{small}} \subset X^n \subset Y^n. \tag{2.11}$$

Corollary 2. *"The open manifold*

$$Y^n \underset{\text{def}}{=} \text{int} \, ((\Delta^3 \times I) \# \infty \# (S^2 \times D^2)), \tag{2.12}$$

which appears in the L.H.S. of (2.2) is geometrically simply-connected at long distance."

Proof. Consider a compact subset $K \subset Y^4$. We can always find an $N < \infty$, such that

$$K \subset (V^4|N) + \left\{ \text{ the 2-handles } \sum_1^N D_i^2 \times \text{int } D_i^2 \right\} \subset Y^4 \tag{2.13}$$

(see (2.2).) Now, since the sort of link $V^4|N$ is smoothly tame, the $Z^4 \underset{\text{def}}{=} (V^4|N) + \sum_1^N D_i^2 \times \text{int } D_i^2$ which appears in (2.13), is the interior of a *compact* bounded geometrically simply-connected manifold X^4. This X^4 can be slightly pulled inside its interior, so as to get the desired engulfing of K by X^4

$$K \subset X^4 \subset Z^4 \subset Y^4.$$

□

Corollary 3. *"For any homotopy 3-sphere Σ^3, the 4-manifold $\Delta^3 \times I$ is geometrically simply-connected at long distance."*

Proof. We define $A_n = (\Delta^3 \times I) \# n \#(S^2 \times D^2)$, and consider the standard inclusions $A_n \subset \operatorname{int} A_{n+1}$. With this, our Y^4 from (2.12) is

$$Y^4 = \overset{\infty}{\underset{0}{\cup}} A_n. \tag{2.14}$$

We consider the compact subset $A_0 = \Delta^3 \times I \subset Y^4$, to which we apply corollary 2; this gies us a compact geometrically simply-connected $X^4 \subset Y^4$, such that

$$A_0 = \Delta^3 \times I \subset X^4 \subset Y^4. \tag{2.15}$$

By compactness, there is a finite n such that

$$A_0 \subset X^4 \subset A_n. \tag{2.16}$$

Since the inclusion $A_0 \subset A_n$ is standard, if we kill the $\# n \#(S^2 \times D^2)$ of A_n with 3-handles, so as to get $\Delta^3 \times I$, the resulting inclusion $A_0 \subset \Delta^3 \times I$ is just our $(\Delta^3 \times I)_{\text{small}} \subset \Delta^3 \times I$, and hence (2.16) yields

$$(\Delta^3 \times I)_{\text{small}} \subset X^4 \subset \Delta^3 \times I. \tag{2.17}$$

$\square$

With corollary 3 above, we have finished the sketchy description of the first part of our program for the Poincaré Conjecture. The next part of the program is, for the time being, only conjectured.

Conjecture 4. *"If $\Delta^3 \times I$ is geometrically simply-connected at long distance, then $\Delta^3 \times I$ is geometrically simply-connected."*

Work by myself and by David Gabai, concerning this conjecture, is in progress. Part of the plan here, involves the quesion of compatibility between the smooth compactification of $V^4|N$ and $V^4|N + 1$, already mentioned.

3. THE STRANGE COMPACTIFICATION THEOREM

Throughout the present section, the conclusion of conjecture 4 will be used as a hypothesis, i.e. we will work with homotopy 3-spheres Σ^3 about which it is **assumed** that $\Delta^3 \times I$ is geometrically simply-connected. But we need a definition, first. Let us consider any smooth manifold pair like (2.5), call it

$$\mathbf{LK} = (B^4, \overset{k<\infty}{\underset{1}{\sum}} S_i^1 \times D_i^2 \subset \partial B^4) \tag{3.1}$$

with null-framing. (This is really what one would like to a **LINK**, as opposed to a pair $\overset{k}{\underset{1}{\sum}} S_i^1 \subset M^3$ which one would call a "classical link.")

Starting with (3.1), we perform the following 2-stage construction:

(3.2.a) We first perform an infinite connected sum along the boundary, far from $\sum_1^k S_i^1 \times D_i^2$

$$B^4 \Longrightarrow B^4 \# \infty \#(S^2 \times D^2).$$

(3.2.b) Then we erase any piece of boundary, except for $\sum_1^k S_i^1 \times \text{int } D_i^2$, which produces the following non-compact 4-manifold (with non-empty boundary)

$$V^4(\mathbf{LK}) \underset{\text{def}}{=} \text{int } (B^4 \# \infty \#(S^2 \times D^2)) \cup \sum_1^k S_i^1 \times \text{int } D_i^2.$$

Notice that while

$$B^4 \# \infty \#(S^2 \times D^2) \tag{3.3}$$

is not a uniquely defined object, as long as the end-point structure is not specified, the $V^4(\mathbf{LK})$, which is called a *stable sort of link*, is unique, once **LK** (3.1) is given.

We can state now the main result of the present section.

Theorem 5. (The strange compactification theorem) *"Let Σ^3 be a homotopy 3-sphere which is such that $\Delta^3 \times I$ is geometrically simply-connected.*

Then there exists a link **LK** *(3.1) with the following two properties.*

A) If we use **LK** *in order to add k handles of index two to B^4, then we get a diffeomorphism*

$$(\Delta^3 \times I)\#(k\#(S^2 \times D^2)) =$$
$$= B^4 + \left\{ \text{the k handles of index two } \sum_1^k D_j^2 \times D_j^2 \text{ defined by } \mathbf{LK} \right\}. \tag{3.4}$$

B) For the stable sort of link $V^4(\mathbf{LK})$ which is attached to our **LK** *we can find a noncompact smooth 4-manifold W^4 with a connected and simply-connected boundary $W^3 = \partial W^4$ and also a smooth embedding*

$$(V^4(\mathbf{LK}), \ \partial V^4(\mathbf{LK})) \xrightarrow{\xi} (W^4, W^3) \tag{3.5}$$

such that:
B-1) The restriction $\xi|\text{int } V^4(\mathbf{LK})$ is a diffeomorphism $\text{int } V^4(\mathbf{LK}) \approx \text{int } W^4$.
B-2) If we consider the core curves

$$\Gamma \underset{\text{def}}{=} \sum_1^k S_i^1 \subset \sum_1^k S_i^1 \times D_i^2 = \partial V^4(\mathbf{LK}), \tag{3.6}$$

then the classical link $(W^3, \xi(\Gamma))$ is trivial."

Before discussing this statement, I will give right away its main application.

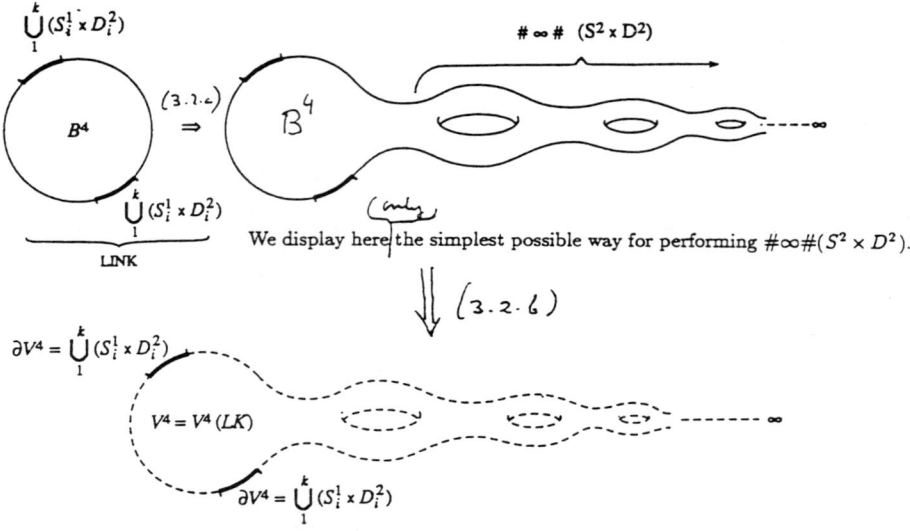

FIG. 2. THE 2-STAGE CONSTRUCTION LEADING TO THE *stable* SORT
OF LINK $V^4(\mathbf{LK})$.

Corollary 6. "*Let Σ^3 be a homotopy 3-sphere which is such that $\Delta^3 \times I$ is geometrically simply-connected. Then, $\Sigma^3 = S^3$.*"

Sketch of Proof. Since $\Delta^3 \times I$ is geometrically simply connected, we have an **LK** (3.1) with the properties A), B) from our theorem above. We also have the classical link $\Gamma \subset S^3 = \partial B^4$ which comes along with (3.1); we denote

$$\pi = \pi_1(S^3 - \Gamma). \tag{3.7}$$

We also consider the fundamenal group at infinity for the non-compact space $V^4(\mathbf{LK})$

$$\pi_1^\infty V^4(\mathbf{LK}) = \varprojlim_C \pi_1(V^4(\mathbf{LK}) - C), \tag{3.8}$$

where the projective limit runs over the compact sets $C \subset V^4(\mathbf{LK})$; it turns out that, with some care, we can handle the base-point problems, and this huge topological *pro-group* is well-defined.

The non-compact space $V^4(\mathbf{LK})$ has two, completely distinct, compactifications, namely

(3.9.1) The *standard compactification* which is, by definition the one-point compactification of (3.3).

(3.9.2) The one-point compactification $W^4 \cup \{\infty\}$, which we call *strange*, provided by theorem 5.

Each of then two compactifications provides a way for computing $\pi_1^\infty V^4(\mathbf{LK})$. If we use the standard compactification, we get

$$\pi_1^\infty V^4(\mathbf{LK}) = \varprojlim(\pi * F_n) \qquad (3.10)$$

when "$*$" means free product and F_n means the free group with n generators. If we use the strange compactification, then we get, on the other hand,

$$\pi_1^\infty V^4(\mathbf{LK}) = \varprojlim F_n. \qquad (3.11)$$

This *pro-free* group is a very complicated topological group, which is *not* free; but it can be shown that every finitely geneated subgroup is free. Since (3.10) combined with (3.11) gives us an injection $\pi \subset \varprojlim F_n$, it follows that π is free. So (according to a well-known theorem of Papakyriakopoulos [Pa]) the classical link (S^3, Γ) is trivial. It follows then from (3.4) that

$$(\Delta^3 \times I) \# k \# (S^2 \times D^2) = B^4 \# k \# (S^2 \times D^2). \qquad (3.12)$$

Going to the respective boundaries and then to the universal coverings, we get an inclusion $\Delta^3 \subset R^3$, which implies that $\sum^3 = S^3$. $\qquad \square$

I will go back now to theorem 5. Extending the boundary $\sum_1^k S_i^1 \times \text{int } D_i^2$ of $V^4(\mathbf{LK})$ to the connected $W^3 = \partial W^4$ involves again an infinite process not unlike the one via which one proves theorem 1. I will not discuss here, at all, the very delicate process of glueing W^3 to the infinity of $V^4(\mathbf{LK})$, but rather talk about the part of the story concerning the pair (W^3, Γ) itself.

The infinite process turns out to provide two "canonical structures" for (W^3, Γ).

I) *The first structure.* The inclusion $\Gamma \subset W^3$ extends to a **PROPER** embedding

$$\sum_1^k (D_i^2 - C_i) \hookrightarrow W^3 \qquad (3.13)$$

where each $C_i \subset \text{int } D_i^2$ is a Cantor set and where

$$\sum_1^k \partial(D_i^2 - C_i) = \Gamma.$$

We will say that Γ *can be pulled to the infinity of* W^3. Here is an easy fact concerning this notion.

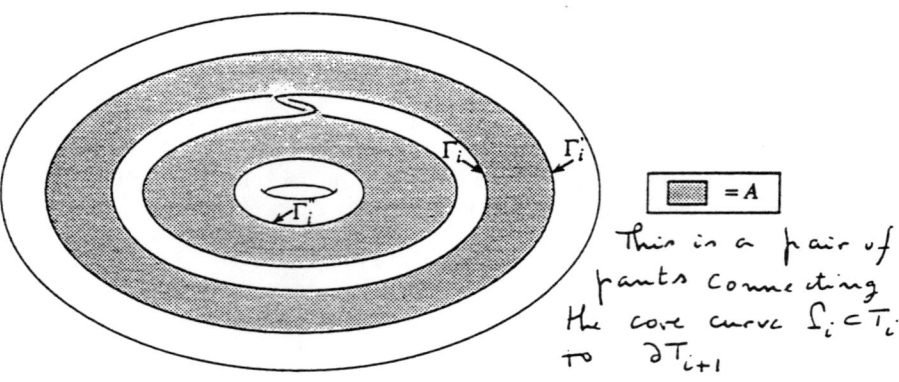

This in a pair of pants connecting the core curve $\Gamma_i \subset T_i$ to ∂T_{i+1}

$\boxed{\blacksquare = A}$

FIG. 3. $T_{i+1} \subset Wh^3$ (see (2.7).)
Here $\Gamma_i \subset \operatorname{int} T_{i+1}$ while $\Gamma'_i + \Gamma''_i \subset \partial T_{i+1}$

Lemma 7. *"Let X^3 be an open simply-connected 3-manifold which is simply-connected at infinity (for which we will use the notation $\pi_1^\infty X^3 = 0$).*

If $\Gamma \subset X^3$ is a classical link which can be pulled at infinity, then (X^3, Γ) is trivial."

Proof. Inside each $D_i^2 - C_i$ we choose a finite system of simple closed curves $\{\Upsilon_i\}$ which split off from $D_i^2 - C_i$ a disk with finitely many holes d_i^2, such that $\partial d_i^2 = \Gamma \cap (D_i^2 - C_i) + \{\Upsilon_i\}$. Since $\pi_1^\infty X^3 = 0$, if we choose $\{\Upsilon_i\}$ sufficiently close to the infinity of X^3, then we can find an extension

$$\begin{array}{ccc} \Gamma \subset \sum_1^k d_i^2 & \longrightarrow & X^3 \\ \downarrow & \nearrow{\scriptstyle g} & \\ \sum_1^k D_i^2 & & \end{array} \tag{3.14}$$

where $\Gamma = \sum_1^k \partial D_i^2$ and where the singular map g has the Dehn property $\Gamma \cap g^{-1} \sum_1^k \operatorname{int} D_i^2 = \phi$. Using Dehn's lemma [Pa], we can produce another extension of $\Gamma \subset X^3$ to an embedded system $\sum_1^k D_i^2 \subset X^3$. $\square$

Without the π_1^∞ hypothesis, lemma 7 no longer holds. For instance, for the Whitehead manifold (2.7), which is well-known to have $\pi_1^\infty Wh^3 \neq 0$, the core of the first torus T_0 can be pulled to infinity (via a D^2-(cantor set)), without the link in quesiton being trivial. One builds up $D^2 - C$ out of pairs of pants like the A in figure 3.

We go back now to (3.13); our W^3, which is built by an infinite process, is a *wild* manifold just like Wh^3, in the sense that

$$\pi_1^\infty W^3 \neq 0. \tag{3.15}$$

So, (3.13) above cannot guarantee us that (W^3, Γ) in trivial. [Incidentally, also, $\pi_2 W^3$ is very large, but we will ignore this fact in the present lecture.]

Fortunately for us, the infinite process produces also a second "canonical structure" for (W^3, Γ).

II) *The second structure.* this is a *lamination* $\mathcal{L}$ of W^3, by planes.

I will start by briefly recalling some basic facts concerning laminations, a notion which is due to W. Thurston [Th1]. So let X^3 be a 3-manifold which is with $\partial X^3 = \emptyset$. A laminination $\mathcal{L}$ of X^3 is defined by a closed subset

$$K = K(\mathcal{L}) \subset X^3 \tag{3.16}$$

such that X^3 admits a smooth atlas $X^3 = \cup_\alpha U_\alpha$ with coordinate charts

$$(U_\alpha, U_\alpha \cap K) = (R^2 \times R, \ R^2 \times \{\text{cantor set}\}). \tag{3.17}$$

This is analogous to a foliation $\mathcal{F}$ of X^3, except that now the transverse structure is a cantor set, instead of being the real line. Like for foliations we can talk about plaquettes $R^2 \times \{x\}$ with $x \in$ Cantor, out of which one builds up connected (2-dimensional) leaves L^2 for $\mathcal{L}$. Again like for foliations, the leaves L^2 have extrinsic topologies induced from $L^2 \subset X^3$ but also *intrinsic* 2-manifold topologies. The connected components of the open subset $X^3 - K(\mathcal{L})$ will be called 3-dimensional leaves of $\mathcal{L}$. Such a 3-dimensional leaf L^3 admits a natural completion $\overline{L}^3$, with $L^3 = \text{int } \overline{L}^3$ and $\partial \overline{L}^3 = \{$the union of L^2, s adjacent to $L^3\}$. For $\overline{L}^3$ one has again to distinguish between extrinsic and intrinsic topology (the only one to be considered here.)

What I have just defined is a lamination *without singulation.* Our lamination $\mathcal{L}$ of W^3 will have singularities which are somehow similar to the 1-prong singularities for measured foliation of surfaces with positive Euler characteristic [Th2], [F-L-P]; they turn out to be relatively benign and we will ignore them here.

It would be wrong to think of laminations as being just another variation on the theme of foliations. Here is an instance of a specific phenomenon for which there is no foliation counterpart. It is possible that all L^2's be contractible without the inclusion map $H_*(L^3) \to H_*(X^3)$ being injective. This turns out to be one of the many obstacles which one has to overcome, in the proof of the strange compactification theorem.

So, for our W^3, the infinite process creates a lamination $\mathcal{L}$ (with mild local singularities). Here is a list of **Properties of the Lamination $\mathcal{L}$ of W^3**.

(3.18.1) Each (non-singular) leaf L^2 of $\mathcal{L}$ is a plane.

(3.18.2) Each 3-dimensional leaf L^3 of $\mathcal{L}$ has

$$\pi_1 L^3 = \pi_1^\infty L^3 = 0.$$

(3.18.3) For the link $\Gamma \subset W^3$ we have

$$\Gamma \cap K(\mathcal{L}) = \emptyset.$$

All three properties above are very good, but the next one is not.

(3.18.4)   The completed 3-dimensional leaves $\overline{L}^3$ of $\mathcal{L}$ can be *wild*, in a sense which I will explain now.

Notice, first, that each connected component $L^2 \subset \partial\overline{L}^3$ has its location at the infinity of L^3 completely determined by a **PROPER** embedded arc

$$[0,1) \xrightarrow{\alpha} L^3,$$

which I call the *wick* of L^2 (inside L^3), and which is defined as follows. Consider a tubular neighbourhood $L^2 \times [0,1] \subset \overline{L}^3$, with $L^2 \times 1 = L^2 \subset \partial\overline{L}^3$ and a point $x \in L^2 = R^2$. Then $\alpha[0,1) = x \times [0,1)$.

The point is that inside the very nice L^3 the wickes can be Artin-Fox type arcs and this is what makes $\overline{L}^3$ wild (and it also accounts for $\pi_1^\infty W^3 \neq 0$.) [Compare this with comment 5) in section 2.] Fortunately, the good properties (3.18.1), (3.18.2), (3.18.3) are enough to save the day, since we have the following

Lemma 8. *1) As a consequence of (3.18.1) to (3.18.3), all the links $(L^3, L^3 \cap \Gamma)$ are trivial.*

2) As an immediate consequence of 1), (W^3, Γ) is also trivial.

The proof is schematized in figure 4.

Here one sees $L^3 \supset \Gamma$ after truncating figure 4.a along the L^2's . One has $\pi_1^\infty L^3 = 0$, and hence via lemma 7, $\Gamma \cap L^3$ is trivial in L^3, and therefore Γ is trivial in W^3 too.

All this should give an idea about how one achieves B-2) in theorem 5.

The proof of theorem 5 is contained in the long series of papers [Po5] and the construction of the good **LK** (3.1) and of $W^4 \supset V^4(\mathbf{LK})$ which these papers give, is too long and intricate to be given here. By contrast, the *description of* $(W^3, \mathcal{L})$ alone is quite short. Here is how it goes.

The infinite process produces (among many other things), the following data.

(3.19.1)   A non-compact 3-manifold T^3, which is an infinite connected sum (along the boundary) of elementary pieces, each of which is either $S^1 \times D^2$ or $S^2 \times I$. There are infinitely many pieces of both kinds and also the end-structure is complicated, (the space of ends is a Cantor set.)

(3.19.2)   Inside ∂T^3 we have an injection

$$\sum_1^\infty \gamma_i \subset \partial T^3,$$

where each γ_i is a simple closed loop, and the set of γ_i's is conjugated to a free basis of $\pi_1 T^3$.

(3.19.3)   There is also a second injection

$$\sum_1^\infty (\Delta_j^2, \partial\Delta_j^2) \subset (T^3, \partial T^3),$$

where the Δ_j^2 is an embedded disk, with int $\Delta_j^2 \subset$ int T^3 and with

$$\partial\Delta_j^2 \cap \gamma_k = \begin{cases} \text{a unique transversal intersection point, if } j = k \quad \text{or} \\ \phi \quad , \quad \text{if } j \neq k. \end{cases}$$

FIG. 4.A. WE SEE HERE W^3, BOUNDARY
OF THE NON COMPACT $W^4 \supset V^4(\mathbf{LK})$.

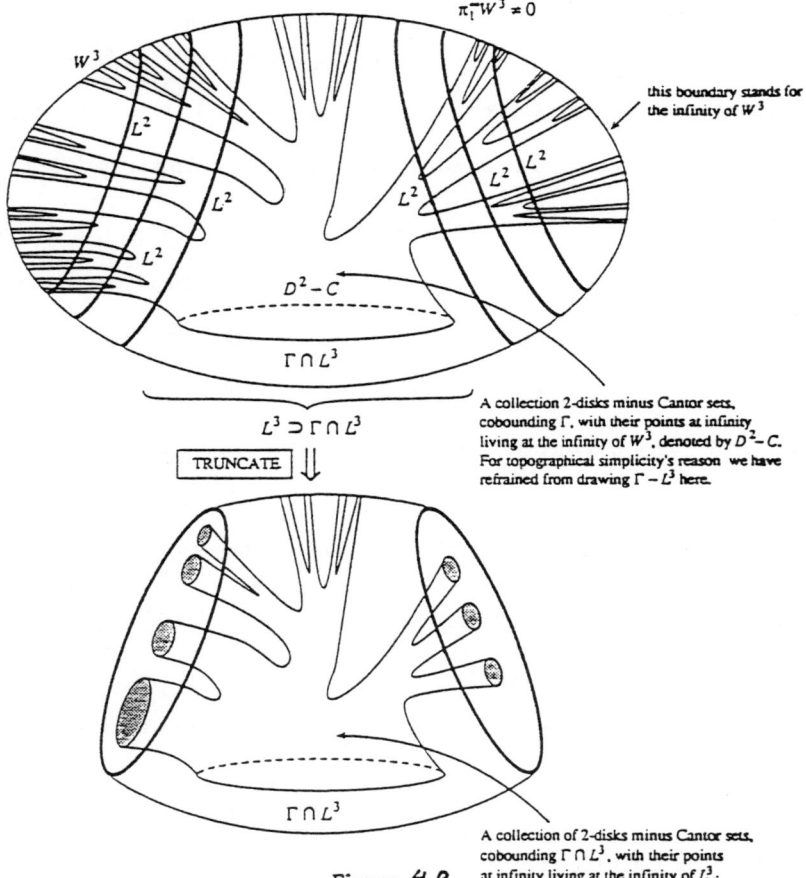

$\pi_1^- W^3 \neq 0$

W^3

L^2

L^2

L^2

L^2

L^2 L^2

L^2

this boundary stands for
the infinity of W^3

$D^2 - C$

$\Gamma \cap L^3$

$L^3 \supset \Gamma \cap L^3$

| TRUNCATE | $\Downarrow$ |

A collection 2-disks minus Cantor sets,
cobounding Γ, with their points at infinity
living at the infinity of W^3, denoted by $D^2 - C$.
For topographical simplicity's reason we have
refrained from drawing $\Gamma - L^3$ here.

$\Gamma \cap L^3$

A collection of 2-disks minus Cantor sets,
cobounding $\Gamma \cap L^3$, with their points
at infinity living at the infinity of L^3.

Figure 4.B

Here one sees $L^3 \supset \Gamma$ after truncating figure 4.a along
the L^2's. One has $\pi_1^\infty L^3 = 0$, and hence via lemma 7,
$\Gamma \cap L^3$ is trivial in L^3, and therefore Γ is trivial in W^3 too.

With this data, we have

$$W^3 = \mathrm{int}\ \{T^3 + [\text{ the infinitely many handles of index} \quad (3.20)$$

$$\lambda = 2, \text{ defined by } \sum_1^\infty \gamma_i \subset \partial T^3]\},$$

and this W^3 comes also equipped with a natural embedding

$$\Gamma \subset \mathrm{int}\ T^3 \subset W^3, \quad (3.21)$$

which is the one from the theorem, and which can be pulled to infinity (3.13).

Notice, before anything else, that *if* the data above would not concern an infinite situation, but a finite one, then the analogue of (3.20) would be just R^3-{a finite set}. But the data (3.19.1) to (3.19.3), as produced by the infinite process, is not only infinite, but also *wild*, in the sense that neither $\sum_1^\infty \gamma_i \subset \partial T^3$ nor $\sum_1^\infty \Delta_j^2 \subset T^3$ are *closed* subsets. [*Exercise*: give a description like (3.20) for the Whitehead manifold Wh^3.]

We would have reached a dead end, if it were not for the following basic fact:

$$\text{Only } \textit{finitely} \text{ many } \Delta_j^2\text{'s touch } \Gamma. \tag{3.22}$$

Achieving (3.22) puts actually enormous constraints on the infinite process which creates all these things.

The lamination $\mathcal{L}$ is defined by taking the limit points of (3.19.3). More specifically

$$K(\mathcal{L}) = \{\text{the set of points } x_\infty \in \text{int } T^3 \tag{3.23}$$

for which there exists an infinite sequence $x_i \in \text{int } \Delta_{n_i}$,

with $n_i \to \infty$ and $\lim x_i = x_\infty\} \subset W^3$.

With this, the crucial property (3.18.3) is an immediate consequence of (3.22).

Remark. It is essential that the $\sum_1^\infty \text{int } \Delta_i \subset \text{int } T^3$ accumulate nicely, making (3.23) a lamination. By contrast, the accumulation pattern of $\sum_1^\infty \Delta_i^2 \subset T^3$ is very wild on ∂T^3. It is actually very hard for a 3-dimensional lamination to exhibit good boundary behaviour.   □

Nothing was said, so far, in this paper, concerning the *infinite processes* which are used in the proofs of theorem 1 and 5. I will end the present section with some comments concerning these processes, in particular for the specific one concerning the smooth tameness theorem.

The initial objects are a certain kind of *representation* for Σ^3, via *singular* objects X^3, of dimension three. For each X^3 there are finitely many possible desingularizations and each of these produces a precise 4-dimensional regular neighbourhood Y^4 of X^3. There is a basic notion of "**coherent**" representation for Σ^3, defined in terms of desingularizations. The explicit definition can be found in any of the references [Ga], [Po1], [Po2], suffices to state here the following:

BASIC FACT 9. "A given Σ^3 admits a coherent representation iff $\Delta^3 \times I$ is geometrically simply-connected." (see [Ga]).

For any Σ^3 there are loads of possible representations (see [Ga], [Po1] where the representations in question are called "collapsible pseudo-sine representation.") Each representation comes equipped with a geometric *square matrix*

$$m = (a_{ij}) \quad , \quad a_{ij} \in Z_+, \tag{3.24}$$

which is, somehow, analogous to the matrices which occur in the context of Markov partitions [Th2], [F-L-P], [Sh] (see also [Sh-Su]). A square matrix like (3.24) can also be thought of as an oriented graph, with

$$S = \text{(set of vertices)} = \text{(set of lines of } m) \approx \text{(set of columns of } m),$$
$$\text{and with } a_{ij} \text{ arrows } i \longrightarrow j.$$
(3.25)

To help understanding what will be coming next, I will state the following

BABY-THEOREM 10. *If for a given Σ^3 there exists a representation such that the oriented graph of m has no closed orbits, then Σ^3 also admits a coherent representation.*

The proof is quite elementary and it can be found in [Ga], [Po4].

So, what the baby-theorem tells us, is that the closed orbits of m are the obstruction in our problem of showing that $\Delta^3 \times I$ is geometrically simply-connected, and the next idea is to push them off to infinity; what we get, is theorem 1.

One has to do this pushing to infinity, first in the purely combinatorial context of geometric square matrices m and/or of oriented graphs. I start by recalling the notion of *block-transformation*. Start with (3.24), (3.25) and with a disjoined partition

$$S = \sum_1^k B_\ell.$$
(3.26)

Next, consider a new geometric square matrix

$$m' = (a'_{\alpha\beta}), \quad \text{with } \alpha, \beta \in \{1, \ldots, k\},$$
(3.27)

where

$$a'_{\alpha\beta} = \sum_{i \in B_\alpha, j \in B_\beta} a_{ij}.$$
(3.28)

The operation $m \to m'$ is a block-transformation.

We consider now any $\sum^3$, some representation and the corresponding m. The purely combinatorial ingredient of the infinite process (which proves theorem 1) is, roughly, the following:

Starting with $m = m_1$ one can build an infinite sequence of *inverse* block transformations, of a very special type

$$m_1 \longleftarrow m_2 \longleftarrow m_3 \longleftarrow \cdots$$
(3.29)

such that

i) There is a limit pattern m_∞ which is an infinite geometric square matrix whose oriented graph has no closed orbits. (Actually, as a matrix, m_∞ is quasi-nilpotent, while in the finite case, "no closed orbits" means nilpotent.)

ii) m_∞ is invariant under further inverse block transformations, of the special type which is considered.

Most important, the combinatorial process above is *geometrically realizable*, in the following sense. There is a machinery called *"honey-comb calculus"* [Ga], [Po2], [Po3] which associates to (3.29) an infinite sequence of representations

$$R_1, R_2, R_3, \ldots$$
(3.30)

of the given Σ^3, with *singular* moves changing $X_n^3 = \{$the singular object of $R_n\}$ into X_{n+1}^3

$$X_1^3 \Longrightarrow X_2^3 \Longrightarrow X_3^3 \Longrightarrow \cdots \tag{3.31}$$

When one goes from X_n^3 to the 4-dimensional smooth regular neighbourhood Y_n^4, (3.31) gives rise to a sequence of smooth embeddings

$$Y_1^4 \hookrightarrow Y_2^4 \hookrightarrow Y_3^4 \hookrightarrow \cdots \tag{3.32}$$

each object in (3.32) being compact and with $\pi_1 \neq 0$. The sort of link V^4 from theorem 1 is $V^4 = \cup_1^\infty Y_n^4$ and it is the quasi-nilpotency of m_∞ which, eventually, leads to the geometric simple connectivity at long distance of $\Delta^3 \times I$.

The proof of theorem 1 is also a paradigm for the proof of theorem 5.

4. INFINITE SIMPLE HOMOTOPY TYPE AND π_1^∞ IN 3-DIMENSIONAL TOPOLOGY

Some of the ideas used in the proofs of theorems 1 and 5 turn out to be useful for other purposes too.

But before I can state the first theorem in this section, I need to come back to the notion of geometric simple connectivity, in the most general context. We start by recalling the basic facts concerning Morse theory on a possibly non-compact manifold with non-empty boundary. We, start with a PROPER smooth function $M^n \xrightarrow{f} R_+$; through-out this lecture "PROPER", written in capital letters, means f^{-1} (compact)= compact, while "proper" with small letters means interior → interior and boundary → boundary. It will be assumed that all the critical points of f are contained int M^n and that they are non-degenerate, and also that the critical points of $f|\partial M^n$ are non-degenerate too. If in the neighbourhood of such a point $p \in \partial M^n$ we have a coordinate chart $(x_1, \ldots, x_n \geq 0)$, the local canonical forms for f fall into two types

$$\text{non-fake type: } f = c - x_1^2 - x_2^2 \cdots - x_\lambda^2 + x_{\lambda+1}^2 + \cdots + x_{n-1}^2 + x_n,$$

$$\text{fake type: } f = c - x_1^2 - x_2^2 - \cdots - x_\lambda^2 + x_{\lambda+1}^2 + \cdots + x_{n-1}^2 - x_n.$$

It is exactly for the non-fake type that there is a change in topology when we go from $f^{-1}[-\infty, c - \epsilon]$ to $f^{-1}[-\infty, c + \epsilon]$. The smooth manifold M^n will be said to be *geometrically simply-connected* if one of the following two equivalent conditions is fulfilled:

1. M^n admits a smooth PROPER Morse function f as above such that for all the critical points of f and for the non-fake critical points of $f|\partial M^n$, the index $\lambda \neq 1$.

2. M^n admits a smooth PROPER handlebody decomposition *without* handles of index $\lambda = 1$ [By "PROPER" handlebody decomposition it is meant that the lateral surface of any given handle is touched only by finitely many attaching zones of other handles. The equivalence of I and II is a standard fact].

Just like in sections 2 and 3 and for more or less analogous reasons we will work again in the **DIFF** category.

The simplest theorem in all the theory under discussion in the following result

Theorem 11. *Let V^3 be an open simply-connected 3-manifold. We consider, for some fixed n, the standard smooth n-ball B^n and the manifold $V^3 \times B^n$, with its canonical smooth structure. Let us assume that we can find a geometrically simply-connected $(n + 3)$-dimensional manifold X^{n+3} endowed with a PROPER smooth embedding $X^{n+3} \subset V^3 \times B^n$ which engulfs the 0-section of $V^3 \times B^n$*

$$V^3 \times (*) \subset X^{n+3} \subset V^3 \times B^n. \tag{4.1}$$

Then, $\pi_1^\infty V^3 = 0$.

We will start by offering some comments concerning this statement. To begin with, it is essential here that V^3 be of dimension 3. If we replace it by V^p, $p > 3$, the corresponding statement becomes false. So, exactly in dimension three, the special case $X^{n+3} = V^3 \times B^n$ of our theorem tells us that π_1^∞ is an obstruction for killing the handles of index one stably. For instance, the 1-handles of the Whitehead manifold Wh^3 cannot be killed, even stably. Notice also that $\pi_1^\infty(V^3 \times R^n) = 0$ (if $n \geq 1$), while $\pi_1^\infty(V^3 \times B^n) = \pi_1^\infty V^3$; the context of our theorem is, really, (open) × (compact). The reader might compare our theorem 1.1 with the following (open) × (open) consequence of J. Stallings' Engulfing theorem [St1]. If W^m is open contractible and if $p > 0$, $m + p \geqslant 5$, then $W^m \times R^p \underset{\text{DIFF}}{=} R^{m+p}$. On the other hand, the failure of the high-dimensional generalization of our result, stems from the following old result of B. Mazur and myself : in any dimension $n \geq 4$ (in particular for $n = 4$, which is the hardest case), there is a contractible compact M^n such that $\pi_1(\partial M^n) \neq 0$ and $M^n \times I = B^{n+1}$ (see [Po10], [Ma]).

One of the peculiarities of dimension 3, in our context, is the following *criterion for simple-connectivity at infinity* in dimension $n = 3$. If V^3 is open simply-connected and if it admits an exhaustion by compact simply-connected sub-complexes

$$K_1 \subset K_2 \subset \cdots \subset V^3, \tag{4.2}$$

then $\pi_1^\infty V^3 = 0$, [Compare this with geometric simple connectivity at long distance.]

The special case $X^{n+3} = V^3 \times B^n$ of theorem 11 is proved in [Po6], [Po7], but the argument given there can be extended so as to prove the full theorem. A far reaching generalization of theorem 11 has been proved by Louis Funar [Fu2].

The techniques from [Po6], [Po7] were obtained by putting together some spare parts of my work on the Poincaré Conjecture. But they have been developed with the aim of investigating the following, well known π_1^∞ CONJECTURE: If M^3 is a closed 3-manifold, then, for its universal covering space $\widetilde{M}^3$, we have $\pi_1^\infty \widetilde{M}^3 = 0$. [I do not know who the author of the conjecture is.]

Let me start with some comments. First, it is easy to see that it would suffice to prove the π_1^∞-conjecture for the special case when $M^3 = K(\pi_1 M^3, 1)$; this would imply the general case too. Now, it so happens that about 1982 M. Davis startled the mathematical community by showing that for every dimension $n \geq 4$ there exists closed manifolds M^n with $M^n = K(\pi_1 M^n, 1)$ (i.e. have contractible $\widetilde{M}^n$) but such that $\pi_1^\infty \widetilde{M}^n \neq 0$ (see [Da]). So, again we are left with a purely 3-dimensional context, just like in the context of theorem 11.

The other remark is that the issue of $\pi_1^\infty \widetilde{M}^3$ only depends on $\pi_1 M^3$, so our question is really a group theoretical problem, in geometrical disguise; but that is something we are used to.

On the other hand, Gromov has taught us that finitely generated groups G are really geometrical objects. If $S = S^{-1}$ is a given finite system of generators for G we can introduce the following norm for $x \in G$:

$\|x\|_S = \{$the minimal length of a word in the alphabet S expressing $x\}$.

This induces a left-invariant distance on G and also on the Cayley graph $\Gamma = \Gamma(G,S)$. From Gromov's viewpoint, $G, \Gamma(G,S)$, or $\widetilde{M^n}$ (where M^n closed, with $\pi_1 M^n = G$) are all equivalent geomtrical objects. On the other hand, there is a fairly long list of *nice geometric conditions* for discrete groups. These conditions include, for instance:

- word hyperbolic (Gromov) [Gr], [G-H], [C-D-P].
- almost convex (This notion, due to J. Cannon [Can], contains word hyperbolic, but also nilpotent; it has been extended in [P-T2] to k-weakly almost convex.)
- automatic (Epstein, Thurston, Cannon,...) [C-E-H-P-T].
- Lipschitz combable (this notion, due to Thurston, extracts the essential feature of "automatic"; its Hausdorff extension is presented in [P-T1]),
- Casson's $C_\alpha, \widehat{C}_\alpha$-conditions [St2], a.s.o

What we can certainly show, for the time being, using techniques like those from [Po6], [Po7], is that, if $\pi_1 M^3$ verifies any reasonable condition of the types above, then $\pi_1^\infty M^3 = 0$. Results along the same lines have also been otained, independently, by Andrew Casson [S-G]. I will give now a sample of precise "nice geometric condition". Let G be a finitely generated group, S a finite system of generators, and $\Gamma = \Gamma(G,S)$ the corresponding Cayley graph.

In Γ we consider the ball of radius n and the sphere of radius n, defined respectively by

$$B(n) = \{x \in \Gamma : \|x\|_S \le n\} \text{ and} \tag{4.3}$$
$$S(n) = \{x \in \Gamma : \|x\|_S \le n\}.$$

Let $k \in Z_+$. We will say that G is k-*almost convex* (with respect to S) if there is a constant $C = C(k) > 0$ such that for any n and any pair $x,y \in S(n)$ with $d_s(x,y) \le k$ there exists a continuous path $L \subset B(n)$ connecting x to y, with $\text{long}_S(L) \le C$. When this condition is fulfilled for every k, our G is, by definition, almost convex (with respect to S). [Notice that, irrespective of k, the geodesic estimate $\text{long}_S(L) = 2n$ always works].

Intuitively, almost convexity means the existence of a lower bound for the "curvature" of $S(n)$, independent of n. It is an open question whether the notion is S-independent like, for instance, the less general Gromov-hyperbolic is.

Theorem 12. *Let M^3 be a closed 3-manifold such that $\pi_1 M^3$ is 3-almost convex (for some S). Then $\pi_1^\infty \widetilde{M^3} = 0$.*

Theorem 12 is proved in [Po8], which also handles the Lipschitz combable case. Other related theorems can be found in [Po9], [P-T1], [P-T2], ...

Now, when dealing with the $\pi_1^\infty V^3$ issue for open manifolds, it will be better to replace the open simply-connected V^3 by a related object: we consider an infinite collection of closed, totally disconnected, tame subsets $K_1, K_2, \dots$ of V^3. By "tame" it is meant here that each K_i is contained inside a smoothly embedded arc $A_i \subset V^3$. Moreover, it will be assumed that $K_i \cap K_j = \emptyset$ (if $i \neq j$) and that any compact set $C \subset V^3$ meets only finitely many K_i's (we will write, symbolically, $K_i \to \infty$). so, we will also assume that $A_i \cap A_j = \emptyset$ and $A_i \to \infty$.

We set $F = \sum_1^\infty K_i$ and introduce the following open 3-manifold

$$V_h^3 \underset{\text{def}}{=} V^3 - F. \tag{4.4}$$

The precise topology of this V_h^3, which we call, generically, "V^3 with very many holes" depends on the topology of F (and it is of course canonically unique when all the K_i's are Cantor sets), but whatever that might be, it is easy to see that $\pi_1^\infty V^3 = \pi_1^\infty V_h^3$; so in our present context, V_h^3 is as good as V^3. I make the following

Conjecture (C). *"Let M^3 be a closed 3-manifold with infinite π_1. There exists an $\widetilde{M}_h^3$ (i.e. a specific $\widetilde{M}^3$ with very many holes) and an $n \in Z^+$, for which we can find a **proper** submanifold $X^{n+3} \subset \widetilde{M}_h^3 \times B^n$ which is geometrically simply-connected and which engulfs the 0-section $\widetilde{M}_h^3 \times 0$."*

In view of Theorem 11 it is clear that conjecture (C) implies the π_1^∞-conjecture. Here is a very first, crude step in the direction of the conjecture (C).

Theorem 13. *"There exists an $\widetilde{M}_h^3$, such that for any integer $n \geq 3$ there is a proper codimension one (stratified) smooth submanifold $\sum = \sum^{n+1} \subset \widetilde{M}_h^3 \times S^{n-1} = \partial\left(\widetilde{M}_h^3 \times B^n\right)$ such that:*

*1) We can find a codimension zero smooth submanifold $X^{n+3} \subset \widetilde{M}_h^3 \times B^n - \sum$ containing $\widetilde{M}_h^3 \times 0$, and which is **PROPER** in $\widetilde{M}_h^3 \times B^n - \sum$.*

2) X^{n+3} is geometrically simply-connected."

The proof, which makes heavy use of [Po11], [Po12] and [Po13], will be given elsewhere. But here are some comments.

A) the stratification of the "**bad locus**" $\sum$ is really of the simplest possible type. There are only strata of codimension zero and one (i.e. of dimension $n + 1$ and n respectively), and along any strata of codimension one we have exactly three strata of codimension zero which meet.

B) Here are some other nice extra features, in the context of Theorem 13:

B.1. The bad locus $\sum$ has various $\pi_1 M^3$-invariance properties, the crudest of which can be expressed as follows. There is a $\pi_1 M^3$-invariant proper stratified submanifold $\sum_1 \subset \widetilde{M}^3 \times B^n$ such that $\sum = \sum_1 \cap \left(\widetilde{M}_h^3 \times B^n\right)$. [One has actually to use here the "correct" embedding $\widetilde{M}_h^3 \subset \widetilde{M}^3$ (see [Po12], which is not exactly the one suggested by (4.4).)

B.2. Our X^{n+3} is the regular neighbourhood of a locally finite simplicial complex. (This is neither true for $\widetilde{M}_h^3$, $\widetilde{M}_h^3 \times B^n$ nor for $\widetilde{M}_h^3 \times B^n - \sum$.)

C) But X^{n+3} is only a **proper** submanifold of $\widetilde{M}_h^3 \times B^n - \sum$ and *not* necessarily of $\widetilde{M}_h^3 \times B^n$ itself. So, our Theorem 13 does not imply conjecture (C) (and hence neither the π_1^∞-conjecture). Actually, without any loss of generality, we can ask that the closure of X^{n+3} in $\widetilde{M}_h^3 \times B^n$ be the full $X \cup \sum$.

D) There is a rather round-about connection between our theorem 13 and hte various results of the general type

$$(\pi_1 M^3 \text{ satisfies some nice geometric condition }) \implies \left(\pi_1^\infty \widetilde{M}^3 = 0\right) \qquad (4.5)$$

of which theorem 12 is an example. Whenever we have a "nice geometric condition" we can apply the machinery which proves Theorem 13, so as to get, additionally, $\sum = \emptyset$, and then we can invoke Theorem 11 in order to conclude that $\pi_1^\infty \widetilde{M}_h^3 = 0$ (and hence that $\pi_1^\infty \widetilde{M}^3 = 0$). Another way to put things is that for the M^3's which have π_1's verifying "nice geometric conditions," **CONJECTURE (C)** is true. I do not offer this as an alternative road to (4.5), it is obviously too complicated, but I do think that in attacking the (C)-conjecture in full generality one is unescapably forced to face the bad locus $\sum$ and to cope with it (see also what is coming next). Actually, the Proof of Theorem 13 also relies heavily on ingredients which apply to *all* simply connected open 3-manifolds, even possibly wild ones and which we will partially review now.

I start by reminding the reader what *weak topology* is all about, in a context useful for our present purposes. We will consider, for some fixed $m \in Z_+$, a sequence of smooth m-manifolds and smooth embeddings

$$Y_1 \subset Y_2 \subset \cdots . \qquad (4.6)$$

Here, for $Y_n \overset{i}{\hookrightarrow} Y_{n+1}$ it is only required that i be injective, with nondegenerate tangent map, but no special resrictions are imposed to $i \,|\, \partial Y_n$, as far as ∂Y_{n+1} is concerned. We define $\varinjlim Y_n$ as being the set $\overset{\infty}{\underset{1}{\cup}} Y_n$ endowed with the following topology:

$$E \subset \cup_1^\infty Y_n \text{ is closed } \leftrightarrow \text{ all } E \cap Y_n\text{'s are closed.}$$

When the ∂Y_n's are non-void this can be a rather weird object. The following example should be useful for understanding our Theorem 14 below.

A Toy Model. We work in $R^2 = (x, y)$ and we start with the lower half plane $A = \{y \le 0\}$.

Let $p_1, p_2, \ldots$ be a bounded monotonly increasing sequence of points on ∂A, converging to some p_∞. Let also $I_n \subset \partial A$ be a small arc centered at p_n. We assume the I_n's two-by-two disjoined, and we consider

$$D_n = \{\text{the half-disk of diameter } I_n \text{ contained in the uper half-plane}\}.$$

The subset $Y_n = A \cup \sum_1^n D_i$, $Y_\infty = A \cup \sum_1^\infty D_i$ of R^2 are submanifolds (we can smoothen them in an obvious way). The $\varinjlim Y_n$ is not a manifold but the obvious "identify" map

$$\varinjlim Y_n \overset{f}{\longrightarrow} Y_\infty$$

is a *continuous bijection*. The inverse map f^{-1} fails to be continuous exactly at p_∞. One should notice that if $q_n \in D_n - I_n$, then, in Y_∞, the sequence $q_1, q_2, \ldots$ is such that $\lim q_n = p_\infty$, while in $\varinjlim Y_n$ the same sequence does not converge at all. Not only is $\varinjlim Y_n$ not a manifold, it isn't even metrizable.

This toy-model should be kept in mind for what comes next. We start now with an arbitrary open simply connected 3-manifold V^3.

Theorem 14. *"There exists a V_h^3 such that for any integer $n \geq 2$ there is a proper codimension one (startified) smooth submanifold*

$$\sum = \sum^{n+1} \subset V_h^3 \times S^{n-1} = \partial(V_h^3 \times B^n) \qquad (4.7)$$

and a sequence of smooth $(n+3)$-dimensional manifolds and smooth embeddings

$$X_1 \subset X_2 \subset \cdots \qquad (4.8)$$

with the following properties:

1) X_1 is geometrically simply-connected and each inclusion $X_i \subset X_{i+1}$ consists of finitely many Whitehead dilatations and additions of handles of index $\lambda \neq 1$.

2) There exist smooth embeddings $X_i \xrightarrow{f_i} V_h^3 \times B^n$ such that the diagram

$$
\begin{array}{ccc}
X_i & \longrightarrow & X_{i+1} \\
& \searrow{}^{f_i} \quad \swarrow{}^{f_{i+1}} & \\
& V_h^3 \times B^n &
\end{array}
\qquad (4.9)
$$

commutes. The map

$$\varinjlim X_i \xrightarrow{f} V_h^3 \times B^n \qquad (4.10)$$

which is induced by the f_i's is a **continuous bijection**.

3) f^{-1} fails to be continuous exactly along $\sum$.

4) $\varinjlim X_i - f^{-1} \sum$ has a natural structure of smooth manifold and the restriction of (4.10) to this set is a diffeomorphism:

$$\varinjlim X_i - f^{-1} \sum \xrightarrow{f} V_h^3 \times B^n - \sum .\text{"} \qquad (4.11)$$

The proof of theorem 14 is contained in [Po12], [Po13] and it uses techniques which are related to those from the proof of the smooth tameness theorem.

Here are some comments concerning theorem 14.

E) In the special case when $V^3 = \widetilde{M}^3$ and $n \geq 3$, the $\sum$ from Theorem 13 is the same as the $\sum$ from Theorem 14 while the X^{n+3} from Theorem 13 is gotten from X_1, after certain manipulations. Remark B.2 above applies not only to X^{n+3} but also to the X_i's.

F) In the general case, we cannot get rid of the bad locus $\sum$, since $\sum = \emptyset$ would imply that $\pi_1^\infty V^3 = 0$ (see Theorem 11.) From this standpoint, Theorem 14 is a best possible result.

G) Generally speaking, the $f_i X_i$'s are not **proper** submanifolds $V_h^3 \times B^n$.

H) As already said, Theorem 14 applies to any open simply connected 3-manifold, even to a wild one like the classical Whitehead manifold W^3 (see [Wh1].) The standard notation for this venerable object, which is closely related to the Casson Handles [Ca1], [Fr1], [F-Q], is actually Wh^3 (see (2.7)), but since we use already the subscript "h", we have changed it here in W^3). At least in this special case of W^3, the game can also be played differently, namely dispensing with the passage (4.4) from W^3 to W_h^3. We can get something which is exactly like the statement of Theorem 14 with "V_h^{3}" (or "W_h^{3}") replaced everywhere by "W^{3}" (*without any* holes), except that the bad locus $\sum$ is now no longer a proper (stratified) submanifold. Instead, $W^3 \times S^{n-1}$ is endowed with a **lamination** $\mathcal{L}$ (see [Th1]) with smooth $(n+1)$-dimensional leaves, the union of which *is* the bad locus $\sum$ (see 3 and 4 in Theorem 14.)

The Cantor set C which appears as the transversal structure to the lamination $\mathcal{L}$ has its own little story. This C is generated by a dynamical feed-back loop which is isomorphic to the one which generates the Julia sets [De] for real quadratic polynomials which are such that the iterates of the unique critical value goes to infinity. So, in some sense, C is a Julia set. All this is explained in detail in [P-T4].

I share with other mathematicians the feeling that wild low-dimensional topology (open 3-manifolds which have $\pi_1^\infty \neq 0$, exotic R^n's and Casson Handles, . . .) should be connected to dynamical chaos.

I)   In the context of (4.10), $f^{-1}\sum$ is also smooth and $f^{-1}\sum \xrightarrow{f} \sum$ is a diffeomorphism, just like (4.11).   □

A tentative project of how one might deal with the (C)-**CONJECTURE** (and hence with the π_1^∞-**CONJECTURE** too), in full generality, is presented in [Po 14]. The inter-relation between $V_h^3 \times B^n$ and $\varinjlim X_i$ is a very crude version of the more sophisticated interplay of structures from [Po 14].

BIBLIOGRAPHY

[B]   Z. Bižaca, *An explicit family of exotic Casson Handles*, (preprint).

[Ca1]   A Casson, *Three lectures on new-infinite constructions in h-dimensional manifold*, in the volume "A la Recherche de la Topologic Perdue" edit. L. Guillon-A. Marin, Birkhäuser (1986).

[Ca2]   A. Casson, *Generalizations and Applications of Block Bundles*, (Cambridge) (1967).

[Can]   J. W. Cannon, *Almost convex groups*, Geom. Dedicata **22** (1987), p. 197–210.

[C-D-P]   M. Coornaert, T. Delzant, A. Papdopoulos, *Géométrie et théoric des Groupes*, S. L. N. 1441 (1980).

[D]   S. Donaldson, *An application of gauge theory to four-dimensional topology*, J. Diff. Geom. 18 (1983) pp. 279-315.

[Da]   M. W. Davis, *Groups generated by reflexions and aspherical manifolds not covered by Euclidean Spaces*, Ann. of Math. **117** (1983), p. 293–324.

[De]   R. Devaney, *An introduction to Chaotic Dynamical Systems*, Addison-Wesley, 1989.

[E-C-H-L-P-T] D.B.A. Epstein, J. W. Cannon, D. F. Holt, S. V. F. Levy, M. S. Patterson and W. Thurston, *Word processing in groups*, James and Bartlett, 1992.

[Fr1]   M. H. Freedman, *The topology of four-dimensional manifolds*, J. Diff. Geom. **17** (1982), p. 357–453.

[Fr2]   M. H. Freedman, *There is no room to spare in four dimensional space*, AMS Notices, 31 (1984) p. 3–6.

[F-Q]   M. H. Freedman and F. Quinn, *The Topology of 4-Manifolds*, Princeton Univ. Press, 1990.

[Fu1]   L. Funar, *TQFT and Whitehead's manifold*, Pripublication de Inst. Fourier (Grenoble) 317 (1995).

[Fu2]   L. Funar, *Infinite simple homotopy and the simple connectivity at infinity of open 3-manifolds*, Preprint (Grenoble) 1996.

[F-L-P]   A. Fathi, F. Landenbach, V. Poénaru (editor), *Travanx de Thurston sur less Surfaces*, Seminain d' Orsay Astérisque, 66–67, 1979.

[Ga]   D. Gabai, *Valentin Poenarú's Program for the Poincaré Conjecture*, in the volume, Geometry, Topology and Physics for Raoul Bott (edited by S. T. Yan) International Pren. (1994) 139–166.

[G]   R. Gampf, *Three exotic R^4's and other anomalies*, J. Diff. Geom., 18 (1983) p. 317–328.

[Gr]   M. Gromov, *Hyperbolic groups*, In "Essays in Group Theorem". (S. M. Gersten, ed.), MSRI Publ. **8** (1987), p. 72–263.

[G-H]   E. Ghys, P. de la Harpe, *Sur les groupes hyperboliques d'apre's*, Mikhael Gromov, Birkhäuser, (1990).

[Q]   F. Quinn, *Lectures on Axiomatic TQFT, in Geometry and Quantum Field Theory*, (edt. Freed – Uhlenbeck), AMS (1995).

[K-S]   R. Kirby, L. Siebenmann, *Foundational essays on topological manifolds, smoothings and triangulation*, Ann. of Math. Studies, 88 (1977).

[Ma]   B. Mazur, *A note on some contractible h-manifolds*, Ann. of Math, 73 (1961) p. 221–228.

[Pa]   C. D. Papakyriakopoulos, *On Dehn's lemma and the asphericity of knots*, Ann. of Math. 66 (1957) p. 1–26.

[Po0]   V. Poénaru, *Infinite processes and the 3-dimensional Poincaré Conjecture*, an outline of the outline preprints. Orsay 89–06 (1989).

[Po1]   ———, *Infinite processes and the 3-dimensional Poincaré Conjecture, I: the collapsible pseudo-spine representation*, Topology, vol. 31, N.3 (1992), 625–656.

[Po2]   ———, *Infinite processes and the 3-dimensional Poincaré Conjecture II: the Honeycomb representation theorem*, (second revised version) Preépublications d'Orsay 93-14 (1994).

[Po3]   ———, *Infinite processes and the 3-dimensional Poincaré Conjecture III : the algorithm*, Prépublications d'Orsay 92-10 (1992).

[Po4]   ———, *Infinite processes and the 3-dimensional Poincaré Conjecture, IV (partie A) : le théorème de non-sauvagerie lisse (the smooth tameness theorem)*, Prépublications d'Orsay 92–83 (1992) (part B) Prépublications d'Orsay 95–33 (1995).

[Po5]   ———, *The Strange Compactification Theorem*, (Part A), IHES Preprint M-95-15 (1995); (Part B), IHES Preprint M-96-43 (1996); Part C is in the process of typing, Parts D, E will follow.

[Po6]   ———, *On the equivalence relation forced by the singularities of a nondegenerate simplicial map*, Duke Math. J. **63** (1991), p. 421–429.

[Po7]   ———, *Killing Handles of Index one stably and π_1^∞*, Duke Math. J. **63** (1991), p. 431–447.

[Po8]   ———, *Almost convex groups, Lipschitz Combings and π_1^∞ for Universal Covering spaces of closed 3-manifolds*, J. Diff. Geom. **35** (1992), p. 103–130.

[Po9]   ———, *Geometry "à la gromov" for the fundamental group of a closed 3-manifold M^3 and the simple connectivity at infinity of $\widetilde{M}^3$*, Topology, **33**.

[Po10]   ———, *Les décompositions de l' hypercube in produit topologique*, Bull. Soc. Math., France 88 (1960) p. 113–129.

[Po11]   ———, *A general finiteness theorem in group theory*, Orsay Preprint 1992.

[Po12]   ———, *Representations of open simply-connected 3-manifolds; a finiteness result*, Orsay Preprint 1992.

[Po13]   ———, *The Handles of Index one of the product of an open simply-connected 3-manifold with a high dimensional ball*, (manuscript).

[Po14]   ———, *π_1^∞ and infinite simple homotopy type in dimension three*, IHES Preprint M-95-6 (1995).

[P-T1]   V. Poénaru and C. Tanasi, *Hausdorff combing of groups...*, Annali Sc. Norm. Sup. Pisa **20** (1993), p. 387–414.

[P-T2]   ———, *k-almost convex groups and $\pi_1^\infty \widetilde{M}^3$*, Geom. Dedicata **48** (1993), p. 57–81.

[P-T3]   ———, *Representations of the Whitehead manifold Wh^3 and Julia sets*, Ann. Toulouse IV, 3 (1995) p. 655–694.

[P-T4]   ———, *Some remarks on Casson Handles*, Preprint (Palermo) 1993.

[S]   J. Stallings, *The piece-wise linear structure of Euclidean space*, Proc. Cambridge Phil. Soc., **58** (1962) p. 481–488.

[S-G]   J. Stallings and S. M. Gersten, *Casson's idea about 3-manifolds whose universal cover is $\mathbb{R}^3$*, Preprint 1991.

[Sh]   M. Shub, *Stabilité globale des systémes dynamicques* Astérisque, **56** (1978).

[Sh-Su]   M. Shub, D. Sullivan, *Homology theory and dynamical systems*, Topology, **14** (1975) p. 109–132.

[Su]   D. Sullivan, *Triangulating Homotopy Equivalences and Homeomorphisms*, Geometric Topology Seminar Notes (Princeton) (1967).

[Ta]   C. H. Taubes, *Gauge theory on asymptotically periodic 4-manifolds*, J. of Diff. Geom., **25** (1987) p. 363–430.

[Th1]   W. P. Thurston, *Three dimensional manifolds, Kleinian groups and hyperbolic geometry*, BAMS, **6** (1982), p. 357–381.

[Th2]   ——————, *On the geometry and dynamics of diffeomorphisms of surfaces*, BAMS, **19** (1988) p. 417–431.

[Wh1]   J. H. C. Whitehead, *A certain open manifold whose groups is unity*, Q. J. of Math **6** (1935), p. 268–279.

[Wh2]   ——————, *On C^1-complexes*, Ann. of Math. **41** (1940), p. 804–824.

ON THE FOUNDATION OF GEOMETRY, ANALYSIS, AND THE DIFFERENTIABLE STRUCTURE FOR MANIFOLDS

Dennis Sullivan

There are levels of structure on a set beneath that of the differentiable structure on a smooth manifold and above that of the topological structure where one has more than enough to define an adequate algebra of differential forms, exterior d and their nonlinear versions connections and curvatures on vector bundles over the space.

One example described in Whitney's book [1] is the structure of any metric gauge which is locally equivalent to that of a polyhedron. By a metric gauge we mean a maximal class of metrics on the set where for any two in the class all the respective local distances $d(x, y)$ are in bounded ratio. The idea of Whitney's construction is to start with Lipschitz chains and put a new norm which makes two chains close if they are up to a small mass error homologous by a chain of small mass. The continuous cochains for this Whitney norm on chains form a graded commutative differential algebra – the Whitney forms for the metric gauge. One obtains in this way forms w so that w and dw have bounded measurable coefficients. Exterior d is a bounded operator for the Whitney norm. These are also forms with square integrable coefficients with exterior d a closable unbounded operator with dense domain containing the Whitney forms.

If the metric gauge is locally equivalent to that of Euclidean space, Teleman [2] developed Hodge theory using these Whitney and L^2 forms. He found that exterior d plus the adjoint of the exterior d relative to any L^2 inner product where multiplication by functions is self adjoint is itself essentially self adjoint.

Such a Hilbert space norm on forms is determined by a bounded measurable Hodge $*$ operator and it is not clear a priori that d and d^* have a common domain. A key point is that d has closed image which follows as usual from deRham theorem. The self adjoint signature operator $d + d^*$ was used by Teleman to develop a version of the Atiyah-Singer Index theorem for these metric gauges. Other corollaries [3, 4] were a construction of characteristic classes and the K-homology orientation [17] from the metric gauge.

One knows the following result [5].

Theorem 1. *Locally Euclidean metric gauges exist and are unique up to small isotopy on every locally Euclidean topological space if the dimension is not equal to four.*

This uses [5] together with the work of Bing and Moise below dimension four and the annulus work of Kirby based on Novikov in dimensions above four. (See [5] for further references.)

The existence and uniqueness result of Theorem 1 is also true for locally Euclidean conformal gauges. By a conformal gauge we mean a maximal class of metrics on a set where for any two metrics in the class all the respective local relative distances $d(x,y)/d(x,y')$ are in a bounded ratio. Now differential forms, wedge product, and exterior d can also be constructed given a locally Euclidean conformal gauge on a set. This is described in [6] and can be based on either work of the Morrey school or that of the Helsinki school. Here one obtains forms that are $p\underline{th}$ power integrable measurable coefficients where (degree of form) × (power of integrability) = (ambient dimension). Exterior d will be an unbounded operator and it is interesting to note that the composition of a local Poincaré lemma singular integral operator with a Sobolev embedding defines a local inverse of exterior d, ($n/k\underline{th}$ integrable k-forms) go by Poincaré lemma transform to ($1\underline{st}$ derivative $n/k\underline{th}$ power integrable $(k-1)$-forms) go by Sobolev embedding to $((n/k-1)\underline{th}$ power integrable $(k-1)$- forms).

Modifications of this picture in dimension zero and the top dimension n are somtimes required; e.g., one can replace bounded measurable functions (L^∞) by functions of mean oscillation (BMO) and replace the integral n-forms (L^1) by the Hardy space n-forms (H^1). This all makes good sense in the locally Euclidean conformal gauge.

In [6] Teleman's work was recapitulated for the locally Euclidean conformal gauge. This phase of the signature operator (but not its absolute value) could be defined in the conformal gauge via an algebraic device that circumvented Teleman's delicate issue of a common domain of d and d^*. This device works in even dimensions and developing the picture in odd dimension presents interesting new features. The Atiyah-Singer theorem was developed also in the conformal gauge and eventually in dimension four the entire Yang-Mills-Donaldson theory, [6].

Then using work of Michael Freedman [7] one can prove the following theorem [6].

Theorem 2. *The conformal gauges of Kaehler complex surfaces in one topological type can form an infinite number of isomorphism classes. Some locally Euclidean topological spaces in dimension 4 do not admit locally Euclidean conformal gauges.*

Using calculations of Donaldson invariants by Friedman and Morgan [8] for Kaehler complex surfaces and the obvious but beautiful fact that two generic algebraic surfaces in a connected algebraic family are diffeomorphic one can deduce the following result.

Theorem 3. *Fixing the conformal gauge of a Kaehler surface determines the diffoemorphism type up to finitely many possibilities.*

Proof. Friedman and Morgan show that two Kaehler surfaces with the same Donaldson invariants up to deformation lie in one algebraic family, [8]. $\square$

Recently, Friedman and Morgan have arrived at a much faster proof of the statement that two Kaehler surfaces with the same Seiberg-Witten invariants (up to deformation) lie in one algebraic family. This swifter calculation does not literally yield a proof of Theorem 3 even if certain conjectures by Mrowka, Kronheimer, and Witten are verified [9]. The point is that the Seiberg-Witten theory depends on the existence of the Dirac operator on spinors ("square roots" of differential forms). *We cojecture [10] that an appropriate Dirac package does not exist for a locally Euclidean conformal gauge or a locally Euclidean metric gauge unless the gauge contains a smooth structure.* To summerize all the above consider the table which is part theorem and part conjecture:

Operator	Structure
phase of signature operator	locally Euclidean conformal gauge
signature operator	locally Euclidean metric gauge
Dirac operator	locally Euclidean differentiable structure

This table was also discovered by Alain Connes in the context of "non-commutative geometry," [11], where the operator on a Hilbert space h plays the primary role in extracting the non-commtative geometry and analysis from the non-commutative topology (a C^* algebra) and its non-commutative measure theory (a self adjoint representation of the algebra in the Hilbert space h). The link is provided by Atiyah's seminal idea relating K-homology and abstract elliptic oprators [12] that the operator and the representation commute modulo lower order terms.

When Connes developed Chern-Weil formalism in the non-commutative context, he discovered cyclic cohomology [11]. These non-commutative ideas come together with [16] in [13] to develop, using the phase of the signature operator, a local formula for the characteristic classes of an even dimensional manifold with a locally Euclidean conformal gauge and a choice of a bounded measurable $*$ in the middle dimension. The discussion is closely related to the measurable Riemann mapping theorem (Morrey-Ahlfors-Bers) in $2D$ and the Donaldson-Yang-Mills theory in $4D$. In a related paper [14] a locally Euclidean metric gauge and a full choice of a bounded measurable $*$ are used to develop a local formula for characteristic classes which converges to the classical Chern-Weil formula when a small parameter tends to zero, at least in regions where the $*$ is smooth. Both these discussions, [13] and [14], can be viewed as an operator theoretic or quantum version of curvature and Chern-Weil formalism.

Let us come to the idea of the differential or smooth structure itself. In [10] one assumes a locally Euclidean metric gauge and makes use of the Whitney forms mentioned above. One can define a "vector bundle of one forms" – a pair (E, γ) consisting of a Lipschitz vector bundle E and a positive bounded embedding γ of its Lipschitz sections into Whitney one-forms [10]. It makes sense to define the torsion of a connection on a vector bundle of one-forms. A cotangent structure for a metric gauge is by definition a vector bundle of one-forms (E, γ) which admits a torsion free connection. The main result of [10] asserts that a cotangent structure on X determines an index function $X \longrightarrow \{1, 2, 3, \cdots\}$ which depends continuously on (E, γ) so that (E, γ) is isomorphic (up to $\varepsilon > 0$) to that of a smooth structure if and only if the index function is identically one, and more generally there are branched covering "charts" whose local degrees agree with the value of the index function.

Thus the notion of cotangent structure provides a way to characterize and generalize the smooth structure inside a locally Euclidean metric gauge.

The connection with Dirac comes from calculations of Cheeger and his former student Chou [15] for polyhedra. Imagine the branching charts of some cotangent structures are equivalent to polyhedral branched covers, then [15] shows the geometric Dirac operator defined away from the branching set is not essentially self adjoint if one takes as common domain the Lipschitz sections of the pulled back spinor bundle by the branched cover. In other words, the Dirac operator "sees" the branching singularities of a cotangent structure by failing to be essentially self adjoint for the natural domain of Lipschitz sections of the relevant spinor bundle.

In setting up the Seiberg-Witten theory one needs abstractly

(1) differential forms and exterior d,

(2) a Hilbert space structure on forms so that multiplication by functions is self adjoint,

(3) a vector bundle of abstract spinors with the expected algebraic relation to forms and a torsion free orthogonal connection with its correct algebraic relation to Clifford multiplication, and finally

(4) the essential self adjointness of the associated Dirac operator with domain the regular sections the spinor bundle.

It seems that (1), (2), and (3) are often possible in the locally Euclidean metric gauge context but that (1), (2), (3), and (4) are only possible in the smooth context [10].

The discussion in the lectures will concentrate on the above remarks related to references [1], [2], [6], [10], and [13].

REFERENCES

1. H. Whitney, *Geometric Integration Theory*, Princeton University Press, Princeton, 1957.
2. N. Teleman, *The index of the signature operator on Lipschitz manifolds*, Inst. Hautes Etudes Sci. Publ. Math. **58** (1983), 39–78.
3. D. Sullivan and N. Teleman, *Analytic proof of Novikov's theorem*, Inst. Hautes Etudes Sci. Publ. Math. **58** (1983), 79–81.
4. N. Teleman, *The index theorem for topological manifolds*, Acta Math. **153** (1984), 117–152.
5. D. Sullivan, *Homeomorphisms and non-Euclidean geometry*, in: Geometric Topology, Proceedings of 1977 Georgia Topology Conference, Academic Press, New York, 1979.
6. S. K. Donaldson and D. P. Sullivan, *Quasiconformal 4-manifolds*, Acta Math. **163** (1989), 181–252.
7. M. Freedman and F. Quinn, *Topology of 4-manifolds*, Princeton University Press, Princeton, 1990.
8. R. Friedman and J. W. Morgan, *Smooth 4-manifolds and Complex Surfaces*, Springer Verlag, 1994.
9. J. W. Morgan, *The Seibert-Witten Equation and Applications to the Topology of Smooth 4-manifolds*, Princeton University Press, Princeton, 1996.
10. D. Sullivan, *Exterior d, the local degree, and smoothability*, in: Proceedings of conference in honor of William Browder, Princeton University Press, 1995.
11. A. Connes, *Non-commutative Geometry*, Academic Press, San Diego, 1994.
12. M. F. Atiyah, *Global theory of elliptic operators*, Proc. of Intl. conf. on Functional Analysis (Tokyo 1969), University of Tokyo Press (1970), 21–30.
13. A. Connes, D. Sullivan, and N. Teleman, *Quasiconformal mappings, operators on Hilbert space, and local formulae for characteristic classes*, Topology **33** (1996), 663–681.
14. H. Moscovici and F. Wu, *The signature operator and local formulae for characteristic classes, finite propagation speed, ...*, J. Functional Analysis (1994).
15. A. W. Chow, *The Dirac operator on pseudomanifolds*, Proc. Amer. Math. Soc. (1984).
16. S. K. Donaldson and P. B. Kronheimer, *The Geometry of Smooth 4-manifolds*, Oxford Clarenda Press, 1990.
17. D. Sullivan, *Geometric periodicity and the invariants of manifolds*, Lecture Notes in Math. 197, Springer, Berlin, 1971, 44–75.

A CONFORMAL INVARIANT
CHARACTERIZING THE STANDARD SPHERE

AUGUSTIN BANYAGA AND JEAN-PIERRE EZIN

ABSTRACT. We define a conformal invariant of compact riemannian manifolds (M, g), and prove that it is non trivial if and only if (M, g) is conformal to the sphere with its standard metric.

1. INTRODUCTION AND STATEMENT OF THE RESULT

Let (M, g) be a riemannian manifold, i.e M is a smooth manifold equipped with a riemannian metric g. The conformal class γ_g of the metric g is the set of all riemannian metrics g' such that $g' = u.g$ for some positive smooth function u on M. The group $G(M, \gamma_g)$ of automorphisms of the conformal class γ_g consists of C^∞ diffeomorphisms φ of M such that $\varphi^* g = \lambda_\varphi.g$ for some positive function λ_φ. This group is known as the **conformal group** of the riemannian manifold (M, g). It contains the groups $I(M, g') = \{\varphi | \varphi^* g' = g'\}$ of isometries of any riemannian metric g' in the conformal class γ_g as non normal subgroups. With the compact-open C^∞ - topology, the groups $G(M, \gamma_g)$ and $I(M, g')$ are finite dimensionnal Lie groups [5].

Two riemannian manifolds $M_i, g_i)$ are said to be conformal if there exists a diffeomorphism $h : M_1 \to M_2$ such that $h^* g_2 = u.g_1$ for some positive function u. If (M_i, g_i) are conformal, then their conformal groups $G(M_i, \gamma_{g_i})$ are isomorphic Lie groups.

The main result of this note is the following

Theorem.

Let (M, g) be a compact and connected riemannian manifold and consider γ_g the conformal class of g. Given $\varphi \in G(M, \gamma_g)$, denote by λ_φ the positive function such that $\varphi^* g = \lambda_\varphi g$. Consider the mapping $D_g : G(M, \gamma_g) \to C^\infty(M)$:

$$D_g(\varphi) = ln(\lambda_{\varphi^{-1}}).$$

Then :

1. The mapping D_g is a 1-cocycle on $G(M, \gamma_g)$ with values into the $G(M, \gamma_g)$ - module $C^\infty(M)$, and its cohomology class in $H^1(G(M, \gamma_g), C^\infty(M))$ is a conformal invariant, we denote $\delta(\gamma_g, M)$.

2. The invariant $\delta(\gamma_g, M) \neq 0$ if and only if (M, g) is conformal to the sphere S^n with its standard metric g_s.

1991 *Mathematics Subject Classification.* 53C12; 53C15.

Key words and phrases. Isometry group, conformal group, conformal class of a riemannian metric, conformal invariants, cohomology of groups.

2. Proofs of the results

The conformal invariant described here is similar to the contact invariant of [1], and the proof of part 1 of the theorem is the same as in [1].

For the general definition of the cohomology groups $H^*(G,K)$ of an abstract group G with values in a G-module K, we refer to [3]. We need just to recall here that the 1-cocycles coincide with the space $Der(G,K)$ of derivations , the coboundaries with the inner derivations $Inder(G,K)$ and

$$H^1(G,K) \approx Der(G,K)/Inder(G,K).$$

The derivations consist of mappings $d : G \to K$ such that $d(g_1 g_2) = g_1.d(g_2) + d(g_1)$ and the inner derivations are those mappings $v : G \to K$ such that there exist $k \in K$ with $v(g) = g.k - k$.

The action of the group $Diff^\infty(M)$ of diffeomorphisms of a smoth manifold on the space of smooth functions $C^\infty(M)$ is given by: $\varphi.f = f \circ \varphi^{-1}$, for $f \in C^\infty(M)$ and $\varphi \in Diff^\infty(M)$.

For $\varphi, \psi \in G(M, \gamma_g)$, we have:

$$((\varphi\psi)^{-1})^* g = (\varphi^{-1})^*(\lambda_{\psi^{-1}} g) = \lambda_{\psi^{-1}} \circ \varphi^{-1}.\lambda_{\varphi^{-1}} g.$$

Hence:

$$\lambda_{(\varphi\psi)^{-1}} = (\lambda_{\psi^{-1}} \circ \varphi^{-1})\lambda_{\varphi^{-1}} \tag{1}$$

Taking the natural logarithm on both sides, using the definition of the action of $G(M, \gamma_g)$ on $C^\infty(M)$, and the definition of the map D_g above we get:

$$D_g(\varphi\psi) = \varphi.D_g(\psi) + D_g(\varphi) \tag{2}$$

This proves that D_g is a cocycle.

If g' is any riemannian metric conformal to g, i.e. $g' = u.g$, and if $\psi \in G(M, \gamma_g)$, then : $D_{g'}(\psi) = ln(\lambda'_{\psi^{-1}})$ where $(\psi^{-1})^*(g') = \lambda'_{\psi^{-1}}.g'$. But

$$(\psi^{-1})^*(g') = (u \circ \psi^{-1}).(\psi^{-1})^* g) = (u \circ \psi^{-1}).\lambda_{\psi^{-1}}.g = ((u \circ \psi^{-1}).\lambda_{\psi^{-1}}/u)g'.$$

Hence $\lambda'_{\psi^{-1}} = \lambda_{\psi^{-1}}.(u \circ \psi^{-1}/u)$. Taking ln on both sides, we get:

$$D_{g'}(\psi) = D_g(\psi) + \psi.m - m \tag{3}$$

where $m = ln(u)$.

This shows that the cohomology class $[D_g] \in H^1(G(M, \gamma_g), C^\infty(M))$ depends only on the conformal class of g, i.e. is a conformal invariant, we denote it by $\delta(\gamma_g, M)$. This proves the statement 1.

The results in statement 2 rely on the following classical results:

For the sphere (S^n, g_s) with its standard metric, it is well known that $G(S^n, \gamma_{g_s})$ is not compact and that $I(S^n, g_s) = O(n+1)$ is strictly contained in $G(S^n, \gamma_{g_s})$ [2].

Myers-Steenrod's theorem [7]. :
If (M,g) is a compact riemannian manifold, then $I(M,g)$ is a compact Lie group.

Lelong-Ferrand's theorem [6].

If a compact riemannian manifold (M, g) is not conformal to the sphere S^n equipped with its standard metric g_s, then the conformal group $G(M, \gamma_g)$ of conformal transformations of (M, g), is compact.

In fact $G(M, \gamma_g)$ is non compact if and only if (M, g) is conformal to (S^n, g_s).

(The converse of Lelong-Ferrand's theorem follows from the trivial proposition below).

Proposition.

Let (M_i, g_i) be conformal, i.e. there exists a diffeomorphism $h : M_1 \to M_2$ such that $h^* g_2 = u.g_1$ for some positive function u. Then :

$$\bar{h} : G(\gamma_{g_1}, M_1) \to G(\gamma_{g_2}, M_2), \ \psi \mapsto h\psi h^{-1}$$

is a homeomorphism , (the groups above are endowed with the compact-open topology).(Actually $\bar{h}$ is a Lie group isomorphisms). In particular if (M, g) is conformal to a sphere with the standard metric, then $G(\gamma_g, M)$ is not compact, and strictly contains $I(M, g)$.

Proof.

It is enough to show that the mapping $\bar{h}$ in the proposition maps $G(\gamma_{g_1}, M_1)$ into $G(\gamma_{g_2}, M_2)$.

Since $h^* g_2 = u.g_1$ and for $\psi \in G(\gamma_{g_1}, M_1)$, we have $\psi^* g_1 = \lambda_\psi g_1$, a direct calculation shows:

$$\bar{h}(\psi)^* g_2 = (((u \circ \psi)/u)\lambda_\psi) \circ h^{-1}.g_2 \tag{4}$$

This formula shows that $\bar{h}$ maps $G(\gamma_{g_1}, M_1)$ into $G(\gamma_{g_2}, M_2)$. Clearly, it is a homeomorphism, for the compact open topology on the considered groups. Therefore , since $G(S^n, \gamma_{g_s})$ is not compact, so is the conformal group of a conformal riemanian manifold (M, g). By Myers-Steenrod theorem, $I(M, g)$ is compact, hence is not eaqual to the non compact conformal group. As a consequence, we get the converse of Lelong-Ferrand's theorem. $\quad\square$

By Lelong-Ferrand's theorem, to prove assertion 2 in the theorem, we only need the following

Lemma.

Let (M, g) be a compact riemannian manifold, then $\delta(\gamma_g, M) = 0$ if and only if $G(M, \gamma_g)$ is compact

Proof.

The vanishing of the invariant $\delta(\gamma_g, M)$ means that there exists a function k on M such that for all $\varphi \in G(M, \gamma_g)$:

$$ln(\lambda_{\varphi^{-1}}) = D_g(\varphi) = k \circ \varphi^{-1} - k$$

Hence

$$(\varphi^{-1})^* g = ((e^{k \circ \varphi^{-1}})/e^k)g$$

Pulling back by φ, we get :

$$g = ((e^{k \circ \varphi^{-1} \circ \varphi})/(e^{k \circ \varphi})).\varphi^* g$$

Which finally reads:
$$e^{-k}g = \varphi^*(e^{-k}.g)$$

This means that $\varphi \in I(M, e^{-k}g)$. We thus have proved that:
$$G(M, \gamma_g) \subset I(M, e^{-k}g) \subset G(M, \gamma_g).$$

Hence $G(M, \gamma_g) = I(M, e^{-k}g)$. By Meyrs-Steenrod theorem, $G(M, \gamma_g)$ is compact.

Conversely suppose $G(M, \gamma_g)$ compact. Denote by μ the Haar measure of $G(M, \gamma_g)$ and let $A = \int_{G(M,\gamma_g)} d\mu$ be its total volume. Consider the following function:

$$\rho = (1/A) \int_{G(M,\gamma_g)} \lambda_\varphi d\mu(\varphi)$$

and the metric
$$\hat{g} = \rho g \in \gamma_g$$

Let $s \in G(M, \gamma_g)$, then :
$${s^{-1}}^* \hat{g} = (\rho \circ s^{-1}){s^{-1}}^* g = (\rho \circ s^{-1})\lambda_{s^{-1}}g = (\rho \circ s^{-1})\lambda_{s^{-1}}(1/\rho)\hat{g}$$

Following [4], we easily see that:
$$(\rho \circ s^{-1})\lambda_{s^{-1}}(1/\rho) = 1 \tag{5}$$

Therefore $ln\lambda_{s^{-1}} = -ln(\rho \circ s^{-1}) + ln\rho$. This gives : $D_g(s) = s.k - k$, where $k = -ln\rho$. Therefore $\delta(\gamma_g, M) = 0$.

For completeness, let us prove (5). Compare [4]. Since
$\rho \circ s^{-1} = (1/A) \int_{G(M,\gamma_g)} \lambda_\varphi \circ s^{-1} d\mu(\varphi)$, and $\lambda_\varphi \circ s^{-1}\lambda_{s^{-1}} = \lambda_{(\varphi \circ s^{-1})}x$, by (1),
we have:

$$(1/\rho)(\rho \circ s^{-1})\lambda_{s^{-1}} = (1/\rho)(1/A) \int_{G(M,\gamma_g)} \lambda_\varphi \circ s^{-1}\lambda_{s^{-1}} d\mu(\varphi)$$

$$= (1/\rho)(1/A) \int_{G(M,\gamma_g)} \lambda_{(\varphi \circ s^{-1})} d\mu(\varphi) = (1/\rho)(1/A) \int_{G(M,\gamma_g)} \lambda_\sigma d\mu(\sigma) = 1$$

since μ is an invariant measure.

The proof of the lemma, and hence the proof of statement 2 of the theorem, are now complete.

$\square$

3. Remarks

1. For the sphere (S^n, g_s), there is a direct argument to prove that $\delta(\gamma_{g_s}, S^n) \neq 0$.

Suppose $\delta(\gamma_{g_s}, S^n) = 0$. This means there exists a smooth function on S^n such that $D_{g_s}(\psi) = f \circ \psi^{-1} - f$ for all ψ in $G(S^n, \gamma_{g_s})$. In particular, if $\psi \in I(S^n, g_s)$ is an isometry, $D_{g_s}(\psi) = 0$. Hence $f \circ \psi^{-1} = f$. The function f is therefore invariant under the group of isometries of S^n. But this group acts transitively on S^n. Hence f is a constant k. This in turn implies that for any ψ in $G(S^n, \gamma_{g_s})$, then $D_{g_s}(\psi) = k \circ \psi^{-1} - k = 0$. This says that any element of $G(S^n, \gamma_{g_s})$ is an isometry.

This is absurd since we know that the isometry group of (S^n, g_s) is strictly smaller that the conformal group[2].

2. Equation 5 means that $G(M, \gamma_g) = I(M, \hat{g})$. In fact the vanishing of the invariant $\delta(\gamma_g, M)$ is equivalent to the fact that the group $G(M, \gamma_g)$ coincides with the isometry group $I(M, g')$ of some conformal metric g'.

Acknowledgement. The first named author would like to thank the Institut de Mathématiques et Sciences Physiques of the University of Benin for its hospitality and support, during the earlier phase of the preparation of this paper.

REFERENCES

1. Banyaga A., *Invariants of contact Structures and transversally oriented foliations*, Ann. Global Analysis and Geometry Vol14 (1996), 427–441.
2. Bourguignon J.P. and Ezin J.P., *Scalar curvature functions in a conformal class of metrics and conformal transformations*, Trans. Amer.Math.Soc. Vol301 ,no 2 (1987), 723–736.
3. Hilton P. and Stamback U., *A course in homological algebra*, Springer Grad.Text in Math (1971).
4. Ishihara S., *Groups of projective transformations and groups of conformal transformations*, J.Math. Soc. Japan Vol9 (1957), 195–227.
5. Kobayashi S. and Nomizu K., *Foundations of Differential Geometry, I*, Interscience Publishers (1963).
6. Lelong-Ferrand J., *Transformations conformes et quasi-conformes des varietes riemanniennes compactes (démonstration de la conjecture de Lichnerowicz)*, Mem.Acad. Royale Belgique Vol39 Fasc 5 (1971).
7. Meyers S.B. and Steenrod N.E., *The group of isometries of a riemannian manifold*, Ann.Math. Vol40 (1939), 400–416.

Augustin Banyaga
Department of Mathematics
The Pennsylvania State University
University Park, PA 16802

Jean Pierre Ezin
Institut de Mathématiques et des Sciences Physiques
National University of Benin
P.O.Box 613, Porto-Novo, Benin

SPACES OF HOLOMORPHIC MAPS FROM $\mathbb{C}P^1$ TO COMPLEX GRASSMANN MANIFOLDS

DAVID E. HURTUBISE

1. INTRODUCTION

In this note we provide a detailed proof of a "well-known folk theorem." This theorem has been used by many authors who study the topology of spaces of holomorphic maps [1] [7] [5]. The theorem gives a description of the space of holomorphic maps from $\mathbb{C}P^1$ to the complex Grassmann manifold $G_{n,n+k}(\mathbb{C})$ in terms of equivalence classes of λ-matrices $M_{n,n+k}(\mathbb{C}[z])$, i.e. $n \times (n+k)$ matrices with entries in the polynomial ring $\mathbb{C}[z]$. The equivalence relation is given by the action of the topological group $GL_n(\mathbb{C}[z])$ consisting of those $n \times n$ λ-matrices whose determinant is a non-zero constant. This group acts on the space of $n \times (n+k)$ λ-matrices by matrix multiplication on the left.

We will show that the action

$$GL_n(\mathbb{C}[z]) \times M_{n,n+k}(\mathbb{C}[z]) \to M_{n,n+k}(\mathbb{C}[z])$$

restricts to an action

$$GL_n(\mathbb{C}[z]) \times P_{n,n+k}(\mathbb{C}[z]) \to P_{n,n+k}(\mathbb{C}[z])$$

where $P_{n,n+k}(\mathbb{C}[z])$ is the space of polynomial maps from $\mathbb{C}$ to the Stiefel manifold $V_{n,n+k}(\mathbb{C})$. The quotient space is in bijective correspondence with the space of holomorphic maps $f : \mathbb{C}P^1 \to G_{n,n+k}(\mathbb{C})$.

$$\text{Hol}(\mathbb{C}P^1, G_{n,n+k}(\mathbb{C})) \longleftrightarrow P_{n,n+k}(\mathbb{C}[z])/GL_n(\mathbb{C}[z])$$

The space of holomorphic maps $f : \mathbb{C}P^1 \to G_{n,n+k}(\mathbb{C})$ of degree d corresponds to the subspace of $P_{n,n+k}(\mathbb{C}[z])/GL_n(\mathbb{C}[z])$ consisting of those matrices such that the determinants of the minors are all polynomials of degree at most d (with at least one determinant having degree d). We will show that when restricted to the space of holomorphic maps of degree d the above bijection is a homeomorphism.

We should note that the fact that a holomorphic map from $\mathbb{C}P^1$ to $G_{n,n+k}(\mathbb{C})$ is *locally* given by a matrix of polynomials follows quickly from Chow's Theorem and the GAGA principal [8] [3]. The theorem proved in this note (without reference to Chow's Theorem or the GAGA principal) improves the local result given by Chow's Theorem.

First, we show that a holomorphic map $f : CP^1 \to G_{n,n+k}(\mathbb{C})$ can be represented by a single *global* matrix of polynomials. Second, we show that the compact open topology on $\mathrm{Hol}(CP^1, G_{n,n+k}(\mathbb{C}))$ agrees with the quotient topology on $P_{n,n+k}(\mathbb{C}[z])/GL_n(\mathbb{C}[z])$ when one restricts to elements of degree d.

The author would like to thank Professor W. Fulton for suggesting the method of proof used in this note.

2. HOLOMORPHIC MAPS AND λ-MATRICES

In this section we show that every holomorphic map $f : CP^1 \to G_{n,n+k}(\mathbb{C})$ can be represented by a λ-matrix. That is, for every holomorphic map $f : CP^1 \to G_{n,n+k}(\mathbb{C})$ there exists a polynomial map $\tilde{f} : \mathbb{C} \to V_{n,n+k}(\mathbb{C})$ such that the following diagram commutes:

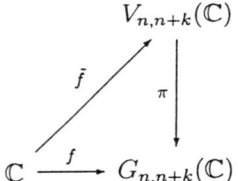

where $CP^1 = \mathbb{C} \cup \infty$.

Let $D^+(z_0) = \{[z_0 : z_1] \in CP^1 | z_0 \neq 0\}$ and $D^+(z_1) = \{[z_0 : z_1] \in CP^1 | z_1 \neq 0\}$.

$$CP^1 = D^+(z_0) \cup D^+(z_1)$$

On $D^+(z_0)$ we have the chart $[z_0 : z_1] \mapsto z_1/z_0$ and on $D^+(z_1)$ we have $[z_0 : z_1] \mapsto z_0/z_1$. In terms of affine coordinates $z = z_1/z_0 \in \mathbb{C}$.

Lemma 1. *Let $\gamma_1^* \to CP^1$ be the tautological holomorphic line bundle. Every holomorphic section s of the m-fold tensor product bundle $\gamma_1^{*\otimes m} \to CP^1$ is a polynomial of degree $\leq m$ in the holomorphic chart on $D^+(z_0)$.*

Proof:

The transition function for $\gamma_1^{*\otimes m}$ from $D^+(z_0)$ to $D^+(z_1)$ is multiplication by $(z_0/z_1)^m$. If we let $z = z_1/z_0$ and identify $D^+(z_0)$ with $\mathbb{C}$ then since s is holomorphic we have,

$$s|_{D^+(z_0)} = \sum_{k \geq 0} a_k z^k$$

and

$$s|_{D^+(z_1)} = \sum_{k \geq 0} b_k z^{-k}.$$

On $D^+(z_0) \cap D^+(z_1)$ we have

$$(z_0/z_1)^m s|_{D^+(z_0)} = s|_{D^+(z_1)}$$

and hence

$$z^{-m} \sum_{k \geq 0} a_k z^k = \sum_{k \geq 0} b_k z^{-k}.$$

Thus,

$$\sum_{k \geq 0} a_k z^k = \sum_{k \geq 0} b_k z^{m-k}$$

for all $z \in \mathbb{C}^*$ and so we must have $a_k = 0$ for $k > m$.

$\square$

The fact that every holomorphic map $f : \mathbb{C}P^1 \to G_{n,n+k}(\mathbb{C})$ can be represented by a λ-matrix follows essentially from the above lemma and the fact that every such holomorphic map is given by the pull-backs under f of $n + k$ sections of the tautological n-plane bundle $\gamma_n^* \to G_{n,n+k}(\mathbb{C})$ which generate the fiber at every point. We give these details first for the case $n = 1$.

The tautological holomorphic line bundle $\gamma_1^* \to \mathbb{C}P^k$ can be defined as the line bundle whose total space is $\mathbb{C}P^{k+1} \backslash \{[0 : \cdots : 0 : 1]\}$ and whose projection map is $p([z_0 : \cdots : z_{k+1}]) = [z_0 : \cdots : z_k]$ (see for instance [3] p. 42). We have an atlas on $\mathbb{C}P^k$ given by the $k + 1$ open sets

$$D^+(z_j) = \{[z_0 : \cdots : z_k] | z_j \neq 0\}$$

for all $j = 0, \ldots, k$ and holomorphic charts $D^+(z_j) \to \mathbb{C}^k$

$$[z_0 : \cdots : z_k] \mapsto (z_0/z_j, \ldots, \widehat{z_j/z_j}, \ldots, z_k/z_j) \in \mathbb{C}^k$$

where the $z_j/z_j = 1$ term is omitted. These charts induce trivializations $h_j : \gamma_1^*|_{D^+(z_j)} \to D^+(z_j) \times \mathbb{C} \to \mathbb{C}$

$$h_j([z_0 : \cdots : z_k : z_{k+1}]) = z_{k+1}/z_j$$

for all $j = 0, \ldots, k$. We have $k + 1$ holomorphic sections of γ_1^* defined by

$$s_j([z_0 : \cdots : z_k]) = [z_0 : \cdots : z_k : z_j]$$

for all $j = 0, \ldots, k$.

Lemma 2. *Let* $f : X \to \mathbb{C}P^k$ *be a continuous map. Then for any trivialization*

$$h : f^*(\gamma_1^*)|_U \to U \times \mathbb{C} \to \mathbb{C}$$

with $x \in U \subseteq X$ *we have*

$$f(x) = [h(s_0^*(x)) : \cdots : h(s_k^*(x))]$$

where s_j^* *is the pull-back of* s_j *along* f *for all* $j = 0, \ldots, k$.

Proof:

If we write $f(x) = [f_0(x) : \cdots : f_k(x)]$, then for any $l = 0, \cdots, k$ we have

$$s_l^*(x) = (x, [f_0(x) : \cdots : f_k(x) : f_l(x)])$$

and in the pull-back of the chart $h_j : \gamma_1^*|_{D^+(z_j)} \to D^+(z_j) \times \mathbb{C} \to \mathbb{C}$ we have

$$h_j^*(s_j^*(x)) = f_l(x)/f_j(x)$$

Thus

$$f(x) = [h_j^*(s_0^*(x)) : \cdots : h_j^*(s_k^*(x))]$$

for all $j = 0, \ldots, k$. For any chart h compatible with h_j^* we have

$$[h(s_0^*(x)) : \cdots : h(s_k^*(x))] = [h_j^*(s_0^*(x)) : \cdots : h_j^*(s_k^*(x))].$$

$\square$

Theorem 3. *Any holomorphic map $f : \mathbb{C}P^1 \to \mathbb{C}P^k$ can be written as*

$$f(z) = [f_0(z) : \cdots : f_k(z)]$$

for all $z \in \mathbb{C}$ where $f_0(z), \ldots, f_k(z)$ are polynomials.

Proof:

This follows from the preceeding Lemma and Lemma 1. In the statement of the theorem $\mathbb{C}P^1 = D^+(z_0) \cup \infty$ and $z \in D^+(z_0) = \mathbb{C}$.

$\square$

For general $n \in \mathbb{N}$ we define the canonical n-plane bundle $\gamma_n \to G_{n,n+k}(\mathbb{C})$ to be the bundle whose total space is

$$\{(p, v)|p \in G_{n,n+k}(\mathbb{C}), v \in p\}.$$

We define the dual of this bundle

$$\gamma_n^* = \text{Hom}(\gamma_n, \mathbb{C})$$

to be the tautological holomorphic n-plane bundle over $G_{n,n+k}(\mathbb{C})$. The reader can check that this definition of γ_n^* agrees with the definition given above when $n = 1$ (see for instance [11] p. 22).

There are $n + k$ canonical holomorphic sections $s_1, \ldots, s_{n+k}$ of γ_n^* defined by

$$s_j(p)[(p, v)] = j\text{th coordinate of } v \in \mathbb{C}^{n+k}$$

for all $j = 1, \ldots, n+k$. These sections generate the fiber of γ_n^* at every point of $p \in G_{n,n+k}(\mathbb{C})$.

The holomorphic coordinate charts on $G_{n,n+k}(\mathbb{C})$ are defined as follows (see for example [4] p. 193). Given an n-plane $p \in G_{n,n+k}(\mathbb{C})$ we begin by choosing any point $\bar{p}$ in the Stiefel manifold $V_{n,n+k}(\mathbb{C})$ above

p. $\tilde{p}$ is an n-tuple of linearly independent vectors in $\mathbb{C}^{n+k}$ which we think of as an $n \times (n+k)$ matrix of complex numbers.

$$\tilde{p} = \begin{pmatrix} a_{11} & a_{12} & \cdots & a_{1\,n+k} \\ a_{21} & a_{22} & \cdots & a_{2\,n+k} \\ \vdots & \vdots & & \vdots \\ a_{n\,1} & a_{n\,2} & \cdots & a_{n\,n+k} \end{pmatrix}$$

We have $G_{n,n+k}(\mathbb{C}) = V_{n,n+k}(\mathbb{C})/GL_n(\mathbb{C})$ where the action of $GL_n(\mathbb{C})$ is given by matrix multiplication on the left, i.e. $\tilde{p} \sim g\tilde{p}$ for all $g \in GL_n(\mathbb{C})$. Since the rows of $\tilde{p}$ are linearly independent there is some minor, say columns $I = (i_1, \ldots, i_n)$, whose determinant is non-zero. By multiplying on the left by the inverse of the minor $\tilde{p}_I$ we get a set of vectors which span the same plane p and whose Ith minor is the identity matrix. The nk entries in the columns not in the Ith minor of $(\tilde{p}_I)^{-1}\tilde{p}$ are local holomorphic coordinates near $p \in G_{n,n+k}(\mathbb{C})$.

Lemma 4. *Let* $f : X \to G_{n,n+k}(\mathbb{C})$ *be a continuous map. Then for any chart*

$$h : f^*(\gamma_n^*)|_U \to U \times \mathbb{C}^n \to \mathbb{C}^n$$

with $x \in U \subseteq X$, $f(x) \subseteq \mathbb{C}^{n+k}$ *is spanned by the rows of the* $n \times (n+k)$ *matrix whose columns are* $h(s_j^*(x)) \in \mathbb{C}^n$ *where* s_j^* *is the pull-back of* s_j *along* f *for all* $j = 1, \ldots, n+k$.

Proof:

Since $s_j^*(x) = (x, s_j(f(x)))$ for all $j = 1, \ldots, n+k$ we need only show that for any chart $\phi : \gamma_n^*|_U \to U \times \mathbb{C}^n \to \mathbb{C}^n$ with $f(x) \in U$ the $n \times (n+k)$ matrix whose columns are $\phi(s_j(f(x))) \in \mathbb{C}^n$ has rows which span $f(x) \in G_{n,n+k}(\mathbb{C})$.

A holomorphic chart around $f(x) \in G_{n,n+k}(\mathbb{C})$ is given by an $n \times (n+k)$ matrix of holomorphic functions whose rows $r_1(p), \ldots, r_n(p)$ span the n-plane $p \in G_{n,n+k}(\mathbb{C})$ for every point p in a neighborhood U of $f(x)$. These row vectors give a basis of the fiber of γ_n above every point $p \in U$ and hence induce a trivialization of $\gamma_n|_U$, i.e. if $(p, v) \in \gamma_n|_U$ satisfies

$$v = \sum_{j=1}^n a_j r_j(p)$$

for some $a_j \in \mathbb{C}$, then the trivialization $\gamma_n|_U \to U \times \mathbb{C}^n \to \mathbb{C}^n$ is defined by $(p, v) \mapsto (a_1, \ldots, a_n)$.

A framing of $\gamma_n^*|_U = \mathrm{Hom}(\gamma_n|_U, \mathbb{C})$ is given by the dual row vectors $r_1^*(p), \ldots, r_n^*(p)$ for all $p \in U$. As for $\gamma_n|_U$ this induces a trivialization of $\gamma_n^*|_U$, i.e. if $(p, v^*) \in \gamma_n^*|_U$ satisfies

$$v^* = \sum_{j=1}^{n} b_j r_j^*(p)$$

for some $b_j \in \mathbb{C}$, then the trivialization $\gamma_n^*|_U \to U \times \mathbb{C}^n \to \mathbb{C}^n$ is defined by $(p, v^*) \mapsto (b_1, \ldots, b_n)$. In this trivialization, the ith component of $s_j(f(x)) \in \mathrm{Hom}(\gamma_n|_{f(x)}, \mathbb{C})$ is

$$s_j(f(x))[(p, r_i(f(x)))] = j\text{th coordinate of } r_i(f(x)) \in \mathbb{C}^{n+k}.$$

Since the lemma holds for this particular trivialization it also holds for any other compatible trivialization.

$\square$

Theorem 5. *For every holomorphic map $f : \mathbb{C}P^1 \to G_{n,n+k}(\mathbb{C})$ there exists a polynomial map $\tilde{f} : \mathbb{C} \to V_{n,n+k}(\mathbb{C})$ such that the following diagram commutes:*

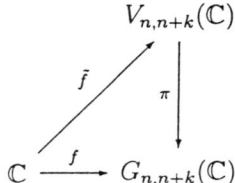

where $\mathbb{C}P^1 = \mathbb{C} \cup \infty$.

Proof:
    The theorem follows immediately from the above lemma, Lemma 1, and the fact that every holomorphic n-plane bundle on $\mathbb{C}P^1$ splits holomorphically into a direct sum of line bundles.

$\square$

3. The Bijection

Let $M_{n,n+k}(\mathbb{C}[z])$ be the set of $n \times (n+k)$ λ-matrices and let $GL_n(\mathbb{C}[z])$ be the set of $n \times n$ λ-matrices whose determinant is in $\mathbb{C}\backslash\{0\}$. $GL_n(\mathbb{C}[z])$ is a topological group that acts on $M_{n,n+k}(\mathbb{C}[z])$ by matrix multiplication on the left.

$$GL_n(\mathbb{C}[z]) \times M_{n,n+k}(\mathbb{C}[z]) \to M_{n,n+k}(\mathbb{C}[z])$$

Note that this action corresponds to polynomial row operations on an element of $M_{n,n+k}(\mathbb{C}[z])$. That is, by multiplying an element of $M_{n,n+k}(\mathbb{C}[z])$ on the left by an element of $GL_n(\mathbb{C}[z])$ we can interchange rows, multiply a row by a non-zero constant, or add a polynomial multiple of one row to another row. (For additional details see [2] Chapter 6.)

Let $P_{n,n+k}(\mathbb{C}[z])$ be the subset of $M_{n,n+k}(\mathbb{C}[z])$ consisting of those matrices whose rows are pointwise linearly independent. That is, those matrices whose rows are in the Stiefel manifold $V_{n,n+k}(\mathbb{C})$ when evaluated at every $z \in \mathbb{C}$. Another way of stating this condition is by requiring that the determinants of the $n \times n$ minors of a matrix in $P_{n,n+k}(\mathbb{C}[z])$ cannot all have a root in common. The space $P_{n,n+k}(\mathbb{C}[z])$ can be identified with the space of polynomial maps from $\mathbb{C}$ to $V_{n,n+k}(\mathbb{C})$. (Compare with Section 3.5 of [10].)

Claim 6. *The action of $GL_n(\mathbb{C}[z])$ on $M_{n,n+k}(\mathbb{C}[z])$ restricts to an action on $P_{n,n+k}(\mathbb{C}[z])$.*

$$GL_n(\mathbb{C}[z]) \times P_{n,n+k}(\mathbb{C}[z]) \to P_{n,n+k}(\mathbb{C}[z]).$$

Proof:

Let $M \in P_{n,n+k}(\mathbb{C}[z])$ and $G \in GL_n(\mathbb{C}[z])$. The determinants of the $n \times n$ minors of GM have the same roots as the determinants of the $n \times n$ minors of M since they differ only by a factor of $\det G \in \mathbb{C}$. This observation proves the claim since an $n \times (n+k)$ matrix of polynomials is in $P_{n,n+k}(\mathbb{C}[z])$ if and only if the determinants of its $n \times n$ minors do not all have a root in common.

$\square$

Theorem 7. *The space of holomorphic maps $f : \mathbb{C}P^1 \to G_{n,n+k}(\mathbb{C})$ is in bijective correspondence with the orbit space of the action of $GL_n(\mathbb{C}[z])$ on $P_{n,n+k}(\mathbb{C}[z])$.*

$$Hol(\mathbb{C}P^1, G_{n,n+n}(\mathbb{C})) \longleftrightarrow P_{n,n+k}(\mathbb{C}[z])/GL_n(\mathbb{C}[z])$$

Proof:

In the previous section we showed that for every holomorphic map $f : \mathbb{C}P^1 \to G_{n,n+k}(\mathbb{C})$ there exists a λ-matrix, say $P \in P_{n,n+k}(\mathbb{C}[z])$, such that $f(z) = \pi(P(z))$ for all $z \in \mathbb{C}$ where $\pi : V_{n,n+k}(\mathbb{C}) \to G_{n,n+k}(\mathbb{C})$ is the map that sends an n-frame to the plane it spans. In order to show that this determines a well-defined element of the orbit space we must show that for any two elements $P_1, P_2 \in P_{n,n+k}(\mathbb{C}[z])$ that satisfy $\pi(P_1(z)) = \pi(P_2(z))$ for all $z \in \mathbb{C}$ there exists $G \in GL_n(\mathbb{C}[z])$ such that $GP_1 = P_2$.

Assume that $\pi(P_1(z)) = \pi(P_2(z))$ for all $z \in \mathbb{C}$. Then there exists a matrix of functions $G(z) = (g_{ij}(z))$ (i.e. $g_{ij} : \mathbb{C} \to \mathbb{C}$ for all $1 \leq i, j \leq n$) such that $G(z)P_1(z) = P_2(z)$ for all $z \in \mathbb{C}$. Since $P_1 \in P_{n,n+k}(\mathbb{C}[z])$ there exists a minor of P_1, say $(P_1)_I$, whose determinant is not the zero polynomial. For every $1 \leq j \leq n$ the jth row of G gives a system of n equations and n unknowns in $g_{1j}, g_{2j}, \ldots, g_{nj}$,

$$(g_{j1}, g_{j2}, \ldots, g_{jn})(P_1)_I = (p_{j1}, p_{j2}, \ldots, p_{jn})$$

where $p_{j1}, p_{j2}, \ldots, p_{jn}$ are the entries in the jth row of the minor $(P_2)_I$. The above is a linear system of n equations and n unknowns over the field of rational functions. Moreover, since the determinant of $(P_1)_I$ is not zero this system of equations has a solution over the field of rational functions. That is, the functions g_{ij} are rational functions which are defined for all $z \in \mathbb{C}$, i.e. the g_{ij} are polynomials. This shows that $G \in GL_n(\mathbb{C}[z])$ since $\det G(z) \neq 0$ for all $z \in \mathbb{C}$. Therefore we have a well defined map

$$\mathrm{Hol}(\mathbb{C}P^1, G_{n,n+n}(\mathbb{C})) \longrightarrow P_{n,n+k}(\mathbb{C}[z])/GL_n(\mathbb{C}[z]).$$

To show that this map has an inverse we need only show that for every orbit there exists a λ-matrix P in the orbit such that the map defined by $f(z) = \pi(P(z))$ is holomorphic for $z \in \mathbb{C}$ and extends continuously to the point at infinity. Then $\infty = [0 : 1] \in \mathbb{C}P^1$ will be a removable singularity and we will have a holomorphic map $f : \mathbb{C}P^1 \to G_{n,n+k}(\mathbb{C})$ defined which clearly corresponds to the orbit of P. If we embed $G_{n,n+k}(\mathbb{C})$ into $\mathbb{C}P^N$ using the Plücker embedding, then the map $f : \mathbb{C} \to G_{n,n+k}(\mathbb{C}) \hookrightarrow \mathbb{C}P^N$ is given by $N + 1$ polynomials and it's clear that a continuous extension to ∞ is simply given by the coefficients of the highest power of z in these $N + 1$ polynomials. Since $G_{n,n+k}(\mathbb{C}) \hookrightarrow \mathbb{C}P^N$ is a closed subset this point must be contained in $G_{n,n+k}(\mathbb{C})$, and hence we have defined a holomorphic map $f : \mathbb{C}P^1 \to G_{n,n+k}(\mathbb{C})$ that satisfies $f(z) = \pi(P(z))$ for all $z \in \mathbb{C}$.

$\square$

4. TOPOLOGICAL ISSUES

The space $\mathrm{Hol}(\mathbb{C}P^1, G_{n,n+k}(\mathbb{C}))$ is given the compact-open topology. Since $G_{n,n+k}(\mathbb{C})$ is a metric space, the compact-open topology on $\mathrm{Hol}(\mathbb{C}P^1, G_{n,n+k}(\mathbb{C}))$ is the same as the topology of compact convergence (see [9] p. 286). Moreover, since $\mathbb{C}P^1$ is compact a sequence of holomorphic maps $f_j \in \mathrm{Hol}(\mathbb{C}P^1, G_{n,n+k}(\mathbb{C}))$ converges to $f \in \mathrm{Hol}(\mathbb{C}P^1, G_{n,n+k}(\mathbb{C}))$ if and only if for every $\epsilon > 0$ there exists $J \in \mathbb{N}$ such that

$$\sup\{d(f_j(z), f(z)) | z \in \mathbb{C}P^1\} < \epsilon$$

for all $j > J$ where d denotes the metric on $G_{n,n+k}(\mathbb{C})$. (For more details see [9] p. 280-283.)

The space $P_{n,n+k}(\mathbb{C}[z])$ is topologized as a subspace of the vector space $\mathbb{C}[z]^{n(n+k)}$, and the orbit space $P_{n,n+k}(\mathbb{C}[z])/GL_n(\mathbb{C}[z])$ is given the quotient topology. The following lemma gives a good intuitive way to understand the topology of $P_{n,n+k}(\mathbb{C})/GL_n(\mathbb{C}[z])$.

Lemma 8. *Let G be a topological group and assume that G acts continuously on a topological space X*

$$G \times X \to X$$

with quotient map $\pi : X \to X/G$. Then π is an open map and a sequence of equivalence classes $\bar{x}_j \in X/G$ converges to $\bar{x} \in X/G$ as $j \to \infty$ if and only if there exists a sequence $x_j \in X$ and an $x \in X$ such that $\pi(x_j) = \bar{x}_j$ for all $j \in \mathbb{N}$, $\pi(x) = \bar{x}$, and $x_j \to x$ as $j \to \infty$.

Proof:

For proof that π is an open map see [6] p. 36. The essential point is that for any open set $U \subseteq X$

$$\pi^{-1}(\pi(U)) = \bigcup_{g \in G} g \cdot U.$$

Now assume that $x_j \to x \in X$ as $j \to \infty$. Since π is continuous we have $\pi(x_j) \to \pi(x)$ as $j \to \infty$. For the other direction assume that we have a sequence $\bar{x}_j \in X/G$, a point $\bar{x} \in X/G$, and an open set U_α containing x_α for each $x_\alpha \in \pi^{-1}(\bar{x})$ such that $\pi^{-1}(\bar{x}_j) \cap U_\alpha = \emptyset$ for all α, j. Then $\pi(\cup_\alpha U_\alpha)$ is an open set containing $\bar{x}$ but not $\bar{x}_j$ for all $j \in \mathbb{N}$. Therefore $\bar{x}_j$ does not converge to $\bar{x}$. $\qquad\square$

To show that the bijection defined in the previous section is a homeomorphism when restricted to maps of a fixed degree, we will reduce the problem to one of maps between projective spaces using the Plücker embedding. Let $N = \binom{n+k}{n}$. The Plücker embedding $Pl : G_{n,n+k}(\mathbb{C}) \to \mathbb{C}P^{N-1}$ is defined by sending a plane to the homogeneous coordinates given by the determinants of the $n \times n$ minors of any element of $V_{n,n+k}(\mathbb{C})$ whose rows span the plane. We have a similar map

$$Pl : P_{n,n+k}(\mathbb{C}[z])/GL_n(\mathbb{C}[z]) \to \mathbb{P}(\mathbb{C}[z]^N)$$

defined by sending an equivalence class $[M]$ to the N-tuple of polynomials (mod $\mathbb{C}^*$) given by the determinants of the $n \times n$ minors of M. This generalized Plücker embedding is well-defined because multiplying M by an element of $GL_n(\mathbb{C}[z])$ can only change the determinants of the $n \times n$ minors of M by an element of $\mathbb{C}^*$.

Lemma 9. $Pl : P_{n,n+k}(\mathbb{C}[z])/GL_n(\mathbb{C}[z]) \to \mathbb{P}(\mathbb{C}[z]^N)$ *is an embedding.*

Proof:

Assume that the determinants of the $n \times n$ minors of $M_1, M_2 \in P_{n,n+k}(\mathbb{C}[z])$ are the same up to multiplication by an element of $\mathbb{C}^*$. Since the standard Plücker embedding $G_{n,n+k}(\mathbb{C}) \to \mathbb{C}P^{N-1}$ is injective, there exists a matrix of functions $G(z) = (g_{ij}(z))$ (i.e. $g_{ij} : \mathbb{C} \to \mathbb{C}$ for all $1 \le i, j \le n$) such that $G(w)M_1(w) = M_2(w)$ for all $w \in \mathbb{C}$. Since $M_1 \in P_{n,n+k}(\mathbb{C}[z])$ there exists a minor of M_1, say $(M_1)_I$, whose determinant is not the zero polynomial. For every $1 \le j \le n$ the jth row of G gives a system of n equations and n unknowns in $g_{1j}, g_{2j}, \dots, g_{nj}$,

$$(g_{j1}, g_{j2}, \dots, g_{jn})(M_1)_I = (l_{j1}, l_{j2}, \dots, l_{jn})$$

where $l_{j1}, l_{j2}, \dots, l_{jn}$ are the entries in the jth row of the minor $(M_2)_I$. The above system of equations is a linear system of n equations and n unknowns over the field of rational functions. Moreover, since the determinant of $(M_1)_I$ is not zero this system of equations has a solution over the field of rational functions. Hence the functions g_{ij} are rational functions that have no poles, i.e. polynomials. Therefore, $G \in GL_n(\mathbb{C}[z])$ and Pl is injective.

The following commutative diagram shows that Pl is continuous.

$$
\begin{array}{ccc}
P_{n,n+k}(\mathbb{C}[z]) & \xrightarrow{\ det \times \cdots \times det\ } & \mathbb{C}[z]^N \\
\downarrow & & \downarrow \\
P_{n,n+k}(\mathbb{C}[z])/GL_n(\mathbb{C}[z]) & \xrightarrow{\ Pl\ } & \mathbb{P}(\mathbb{C}[z]^N)
\end{array}
$$

To see that the inverse map is continuous it suffices to show that the composite

$$
\begin{array}{c}
P_{n,n+k}(\mathbb{C}[z]) \\
\ \ \downarrow \pi \\
P_{n,n+k}(\mathbb{C}[z])/GL_n(\mathbb{C}[z]) \xrightarrow{\ Pl\ } \mathbb{P}(\mathbb{C}[z]^N)
\end{array}
$$

maps open sets to open sets in its image.

Every point $M \in P_{n,n+k}(\mathbb{C}[z])$ has an open neighborhood given by perturbing the coefficients of the entries of M by $\pm\epsilon$ which maps onto an open neighborhood of $(Pl \circ \pi)(M)$. That is, $(Pl \circ \pi)(M) \in \mathbb{P}(\mathbb{C}[z]^N)$ has homogeneous coordinates which are linear functions in the coefficients of the polynomial entries of M. Since a linear function of several variables is an open map $Pl \circ \pi$ is an open map.

$\square$

Theorem 10. *The map*

$$\phi : Hol(\mathbb{C}P^1, G_{n,n+k}(\mathbb{C})) \longrightarrow P_{n,n+k}(\mathbb{C}[z])/GL_n(\mathbb{C}[z])$$

which sends a holomorphic map to the equivalence class of the λ-matrix P such that $f(z) = \pi(P(z))$ for all $z \in \mathbb{C}$ is continuous. When ϕ is restricted to maps of a fixed degree it is a homeomorphism onto its image.

Proof:

Let $N = \binom{n+k}{n}$. The following diagram commutes.

$$
\begin{array}{ccc}
\mathrm{Hol}(\mathbb{C}P^1, G_{n,n+k}(\mathbb{C})) & \xrightarrow{\ \phi\ } & P_{n,n+k}(\mathbb{C}[z])/GL_n(\mathbb{C}[z]) \\
\Big\downarrow{\scriptstyle \circ Pl} & & \Big\downarrow{\scriptstyle Pl} \\
\mathrm{Hol}(\mathbb{C}P^1, \mathbb{C}P^N) & \xrightarrow{\ \phi\ } & \mathbb{P}(\mathbb{C}[z]^N)
\end{array}
$$

Hence by the preceeding lemma it suffices to prove the theorem for the case $n = 1$ since the restriction of a continuous map is continuous.

$\mathrm{Hol}(\mathbb{C}P^1, \mathbb{C}P^N)$ has countably many components. The components are distinguished by the topological degrees of the maps. So it suffices to show that a sequence $f_j \in \mathrm{Hol}(\mathbb{C}P^1, \mathbb{C}P^N)$ of fixed degree d converges to $f \in \mathrm{Hol}(\mathbb{C}P^1, \mathbb{C}P^N)$ if and only if $\phi(f_j) \to \phi(f)$ as $j \to \infty$. Suppose that in homogeneous coordinates

$$f_j(z) = (p_j^0(z) : p_j^1(z) : \cdots : p_j^N(z))$$

and

$$f(z) = (p^0(z) : p^1(z) : \cdots : p^N(z)).$$

Since f_j and f are of degree d for all j we may assume that $p_j^0(z)$ and $p^0(z)$ are monic polynomials of degree d for all j. This means that $\phi(f_j)$ and $\phi(f)$ all lie in a single coordinate chart of $\mathbb{P}(\mathbb{C}[z]^N)$. Hence $\phi(f_j)$ converges to $\phi(f)$ if and only if for all $0 \le i \le N$ the coefficients of $p_j^i(z)$ converge to the coefficients of $p^i(z)$ as $j \to \infty$.

Assume that $f_j(z)$ converges to $f(z)$ uniformly for all $z \in \mathbb{C}P^1$. This implies that for every $0 \le i \le N$ and for a generic $z \in \mathbb{C}$ (where $p_j^0(z) \ne 0$ and $p^0(z) \ne 0$)

$$\lim_{j \to \infty} \frac{p_j^i(z)}{p_j^0(z)} = \frac{p^i(z)}{p^0(z)}$$

Therefore the coefficients of $p_j^i(z)$ converge to the coefficients of $p^i(z)$ for all $0 \le i \le N$.

Now assume that for all $0 \leq i \leq N$ the coefficients of $p_j^i(z)$ converge to the coefficients of $p^i(z)$. We want to show that for every $\epsilon > 0$ there exists a J such that for all $j > J$

$$\sup\{d(f_j(z), f(z))|z \in \mathbb{C}P^1\} < \epsilon$$

where $d(-,-)$ denotes the metric on $\mathbb{C}P^N$. There are several ways of describing this metric. One way is to take the angle between two lines in $\mathbb{C}^{N+1}$ as the metric. An equivalent choice is to take the Hermitian inner product of a unit vector in the first line with a unit vector in the orthogonal complement of the second line. For instance:

$$\frac{\sum_{i=0}^{(N-1)/2} p_j^{2i}(z)\overline{p^{2i+1}(z)} - \overline{p_j^{2i+1}(z)}p^{2i}(z)}{||(p_j^0(z) : p_j^1(z) : \cdots : p_j^N(z))|| \, ||(p^0(z) : p^1(z) : \cdots : p^N(z))||}$$

(Here we have assumed that N is odd. If N is even, then embed $\mathbb{C}P^N$ in $\mathbb{C}P^{N+1}$ by taking the last coordinate to be zero.)

Pick any $\epsilon > 0$. For any closed disk $D(r)$ of radius r the above expression shows that for all $z \in D(r)$ there exists a J_1 such that for all $j > J_1$ we have $d(f_j(z), f(z)) < \epsilon/2$. If r is large then the polynomials will behave like their highest terms when $|z| > r$, and hence it is possible to pick a J_2 such that for all $j > J_2$ we have $d(f_j(z), f(z)) < \epsilon/2$ for all $|z| > r$. Taking $J = \max\{J_1, J_2\}$ we see that $f_j(z)$ converges to $f(z)$ uniformly for all $z \in \mathbb{C}P^1$.

$\square$

References

[1] F.R. Cohen, R.L. Cohen, B.M. Mann, and R.J. Milgram. The topology of rational functions and divisors of surfaces. *Acta Math.*, 166:163–221, 1991.

[2] F.R. Gantmacher. *The Theory of Matricies Vol. 1.* Chelsea Publishing Company, New York, 1960.

[3] P. Griffiths and J. Adams. *Topics in Algebraic and Analytic Geometry.* Princeton University Press, Princeton, 1974.

[4] P. Griffiths and J. Harris. *Principles of Algebraic Geometry.* John Wiley and Sons, New York, 1978.

[5] D.E. Hurtubise and M.D. Sanders. Compactified spaces of holomorphic curves in complex Grassmann manifolds. *Preprint*, 1997.

[6] K. Kawakubo. *The Theory of Transformation Groups.* Oxford University Press, Oxford, 1991.

[7] B. Mann and R.J. Milgram. Some spaces of holomorphic maps to complex Grassmann manifolds. *J. Diff. Geom.*, 33:301–324, 1991.

[8] D. Mumford. *Algebraic Geometry I: Complex Projective Varieties.* Springer-Verlag, New York, 1995.

[9] J.R. Munkres. *Topology.* Prentice-Hall, Englewood Cliffs, New Jersey, 1975.

[10] A. Pressley and G. Segal. *Loop Groups.* Clarendon Press, Oxford, 1990.

[11] R.O. Wells. *Differential Analysis on Complex Manifolds*. Springer-Verlag, New York, 1980.

DEPARTMENT OF MATHEMATICS, PENN STATE UNIVERSITY – THE ALTOONA COLLEGE, 101B EICHE, ALTOONA, PA 16601-3760
E-mail address: hurtubis@math.psu.edu

SETS WITH LIE ISOMETRY GROUPS

H. MOVAHEDI-LANKARANI AND R. WELLS

ABSTRACT. The following result is established.

Theorem: Let μ be a geometric measure supported on a compact subset X of a Euclidean space and let d be a smooth metric on X. Then the group $\mathcal{G}$ of isometries of (X, d) is a (finite dimensional) Lie group and the action $\mathcal{G} \times X \longrightarrow X$ is C^1.

The main tool used to prove this result is the embedding $\iota_p \colon X \longrightarrow L^p(\mu)$ defined by setting $\iota_p(x) = d(x, \ )$.

Theorem: For (X, d, μ) as above and $1 \leq p < \mathrm{Dim}(X)$, the map ι_p is a C^1 embedding.

Here, $\mathrm{Dim}(X)$ is the pointwise lower scaling dimension and the theorem generalizes to certain natural subsets X of Hilbert space. Pathological subsets of Hilbert space (with respect to the Rademacher Differentiability Theorem) may be constructed by using the following result.

Theorem: For (X, d) an ultrametric space, μ a D-dimensional geometric measure on (X, d), and $0 < s = 1 - D/p$ with $1 \leq p < \infty$, the map $\iota_p^s \colon X \longrightarrow L^p(\mu)$ is a locally bi-Lipschitz embedding.

1. INTRODUCTION AND MAIN RESULT

Of course it is well known that the isometry group of a Riemannian manifold is a Lie group. The purpose of this paper is to prove the following theorem; and in particular, we give a new proof of the classical theorem.

Main Theorem. *Let μ be a geometric measure supported on a compact subset X of a Euclidean space and let d be a smooth metric on X. Then the group $\mathcal{G}$ of isometries of (X, d) is a (finite dimensional) Lie group and the action $\mathcal{G} \times X \longrightarrow X$ is C^1.*

Here, a measure μ on X is said to be *geometric* of dimension $D \geq 0$ (a real number) if there exist $0 < a < A$ and $0 < r_0$ such that

$$a\, r^D \leq \mu\left(B(x, r)\right) \leq A\, r^D$$

for all $x \in X$ and $0 \leq r \leq r_0$, where $B(x, r)$ denotes the closed ball in X with radius r centered at x. To define a smooth metric, we say that a C^1 function on X is simply the restriction to X of an ordinary C^1 function on its ambient Euclidean space [14, 9]. Then by a *smooth metric* on X we mean any metric d such that the canonical map $\iota \colon X \longrightarrow L^2(\mu)$ given by $\iota(x) = d(x, \ )$ is a C^1 immersion. In particular, any metric d is smooth if the map $(x, y) \mapsto d(x, y)^r$ is C^1 for some $r > 1$.

1991 *Mathematics Subject Classification.* Primary: 57R55, 57R40, 58A05; Secondary: 58D19.

Key words and phrases. smooth manifold, smooth structure, C^1 structure, geometric measure, Riemannian metric, dimension, ultrametric, Lie group, quotients.

The second author was partially supported through grant N00014-90-J-4012 of the Office for Naval Research.

Consequently, the arclength metric defined by a Riemann metric on $\mathbb{R}^n$ is smooth, and so is any restriction.

The main tool in our proof is this canonical map, which differs essentially from the classical one $\iota: X \longrightarrow C(X) \subset L^\infty(\mu)$ (see [2, 5]) in that the classical map is nowhere C^1 even though it is an isometric embedding. For example, $\iota(S^1) \subset C(X)$ has a "corner" at every point.

Finally, we note that arclength metrics on C^1–path connected subsets of $\mathbb{R}^n$ are not necessarily smooth in our sense. There are such subsets of $\mathbb{R}^3$ supporting infinite dimensional isometry groups – and Lie subgroups of arbitrarily large dimension. (See the last section.)

2. CANONICAL EMBEDDINGS

For convenience, in this section we develope the canonical embedding ι_p in the context of the C^1 category. Following Nomizu [14] and Milnor [9] one obtains a modest extension of the C^1 category by allowing the objects to be pairs $(X, \mathcal{F})$, where X is a closed subset of some Banach space $\mathbb{B}$ and $\mathcal{F}$ is the restriction to X of the classical $C^1(\mathbb{B})$. The C^1 maps from one object to another are those respecting the the C^1 functions. By a C^1 diffeomorphism we mean an isomorphism in this category. It is clear how subsets define subobjects and accordingly a C^1 embedding is a diffeomorphism onto a subobject. (Of course, a C^r category may be defined in exactly the same way.)

We say that a subset X of a Banach space is *spherically compact* if the *norm* closure of the set $\{(x - y)/\|x - y\| \mid x, y \in X \text{ with } x \neq y\}$ is *norm* compact [8, 11]. Although not obviously so, spherical compactness is an invariant of the extended C^1 category. The following theorem is the main result of this section.

Theorem 2.1. *Let X be a compact and spherically compact subset of a Hilbert space $\mathbb{H}$. Let μ be a finite regular Borel measure on $\mathbb{H}$ with X the closed support of μ and suppose that $1 \leq p < \inf_{x \in X} \underline{D}(x) < \infty$. Then the map $\iota_p: \mathbb{H} \longrightarrow L^p(\mu)$ given by $\iota_p(x) = \|x - \ \|_p$ is C^1 and restricts to a bi-Lipschitz embedding of X.*

Here, the *lower scaling dimension* $\underline{D}(x)$ of (X, d, μ) at the point $x \in X$ is defined by setting

$$\underline{D}(x) = \liminf_{r \to 0} \frac{\log \mu(B(x, r))}{\log r} .$$

Of course the map $(\iota_p|X)^{-1}$ need not be C^1, but as a consequence, we see that when $\mathbb{H}$ is finite dimensional, we do recover the C^1 structure of X.

Corollary 2.2. *With the same hypotheses as in Theorem 2.1, if $\mathbb{H}$ is finite dimensional, then the map $\iota_p: X \longrightarrow L^p(\mu)$ is a C^1 embedding.*

The rest of this section is devoted to the proof of the above two results. We begin with the following lemma.

Lemma 2.3. *Let μ be a finite regular Borel measure supported on a subset X of a Hilbert space $\mathbb{H}$ and suppose that there exist positive constants A, D, and r_0 such that $\mu(B(x, r)) \leq A r^D$ for all $x \in X$ and $0 \leq r \leq r_0$. If $p < D$, then there exists $K < \infty$, independent of x, such that the p-potential at x*

$$\int_{\mathbb{H}} \frac{d\mu(z)}{\|x - z\|^p} < K$$

for all $x \in \mathbb{H}$.

Proof. (cf., [3, pp. 78–79]) First observe that if $\mu(B(x,r)) \leq Ar^D$ for all $x \in X$, then $\mu(B(x,r)) \leq Ar^D$ for all $x \in \overline{X}$. Hence, we may as well assume that X is the closed support of μ and prove the conclusion for $x \in X$. Let $\varepsilon = (D - p)/2$ and for $n \geq 1$ let $r_n = (1/n)^{1/\varepsilon}$. For each $n \geq 1$ define

$$f_n(z) = \begin{cases} 0 & \text{if } \|x - z\| < r_n \\ \dfrac{1}{\|x - z\|} & \text{if } \|x - z\| \geq r_n \end{cases}$$

and set $F_n = \int_{\mathbb{H}} f_n^p d\mu$. Then for $n \geq 1$ we have

$$0 \leq F_{n+1} - F_n = \int_{r_{n+1} \leq \|x - z\| \leq r_n} \frac{d\mu(z)}{\|x - z\|^p} \leq \frac{1}{r_{n+1}^p} \mu[B(x, r_n) \setminus B(x, r_{n+1})]$$

$$\leq A\left(\frac{r_n}{r_{n+1}}\right)^p r_n^{D-p} \leq A\, 2^{p/\varepsilon} \left(\frac{1}{n}\right)^{(D-p)/\varepsilon} .$$

Hence, we have

$$\int_{\mathbb{H}} \frac{d\mu(z)}{\|x - z\|^p} = F_1 + \sum_{n=1}^{\infty} (F_{n+1} - F_n)$$

$$\leq F_1 + A\, 2^{p/\varepsilon} \sum_{n=1}^{\infty} \frac{1}{n^2} < \infty .$$

Of course as it stands, F_1 may depend on x, but there is an obvious upper estimate independent of x. $\square$

Lemma 2.4. *Let X, $\mathbb{H}$, and μ be as in* Lemma 2.3 *and let $1 \leq p < \inf_{x \in X} \underline{D}(x) < \infty$. Then the map $\iota_p \colon \mathbb{H} \longrightarrow L^P(\mu)$ defined by setting $\iota_p(x) = \|x - \ \|$ is C^1.*

Proof. For each $x \in X$, we define a map $l_x \colon \mathbb{H} \longrightarrow L^P(\mu)$ by setting

$$(l_x \xi)(z) = \begin{cases} \dfrac{\langle \xi, x - z \rangle}{\|x - z\|} & \text{for } z \neq x \\ 0 & \text{for } z = x, \end{cases}$$

where $\langle\ ,\ \rangle$ denotes the inner product in $\mathbb{H}$. We first show that ι_p is differentiable at each $x \in \mathbb{H}$ with $d\iota_p(x) = l_x$. To this end, let $\varepsilon > 0$ be given. If $z \neq x$, then we have

$$\|\iota_p(x + \xi) - \iota_p(x) - l_x \xi\|_p^p \leq 2^{p+1} \left[\int_{\mathbb{H}} \frac{\|\xi\|^{2p}}{\|x - z\|^p} d\mu(z) \right] \leq 2^{p+1} K \|\xi\|^{2p}$$

by Lemma 2.3. Hence, $\|\iota_p(x + \xi) - \iota_p(x) - l_x \xi\|_p \leq \varepsilon \|\xi\|$ whenever we have $\|\xi\| \leq \varepsilon\, 2^{-(p+1)/p} K^{-1/p}$.

Next, in order to show that ι_p is C^1, let $x, y \in \mathbb{H}$. Then for $z \in X$, with $z \neq x$ and $z \neq y$, and $\xi \in \mathbb{H}$ we have

$$|(d\iota_p(x)\xi)(z) - (d\iota_p(y)\xi)(z)| \leq \frac{2\|\xi\|\,\|x - y\|}{\|x - z\|} .$$

Again, by Lemma 2.3, we obtain

$$\|d\iota_p(x)\xi - d\iota_p(y)\xi\|_p \le 2\,\|\xi\|\,\|x-y\| \left(\int_X \frac{d\mu(z)}{\|x-z\|^p} \right)^{1/p} \le 2\,\|\xi\|\,\|x-y\|\,K^{1/p}\,.$$

By taking sup over $\|\xi\| \le 1$ we see that the operator norm $\|d\iota_p(x) - d\iota_p(y)\| < \varepsilon$ for $\|x-y\|$ sufficiently small, and the proof is complete. $\square$

Following [4] and [8], for a subset X of a Banach space $\mathbb{B}$, we define the $0^{\underline{\text{th}}}$ level tangent cone of X at $x \in X$ by setting

$$C_x^0 X = \text{the set of limits of the form } \lim_{n\to\infty} (y_n - z_n)/\|y_n - z_n\|,$$

where $\{y_n\}_{n\ge 1}$ and $\{z_n\}_{n\ge 1}$, with $y_n \ne z_n$, are sequences in X
converging to x in norm.

Now we are ready to prove Theorem 2.1.

Proof of Theorem 2.1. Since X is the closed support of μ, it is clear that $\mu(B(x,r)) > 0$ for all $x \in X$ and $r > 0$. Recall that by Lemma 2.4, the map ι_p is C^1 with $(d\iota_p(x)\xi)(z) = \langle \xi, x - z \rangle / \|x - z\|$. Let $x \in X$ and let $z_o \in X \setminus \{x\}$ with $(d\iota_p(x)\xi)(z_o) \ne 0$. Then for some $r > 0$ we have $(d\iota_p(x)\xi)(z) \ne 0$ for all $z \in B(z_o, r)$ and so the function $d\iota_p(x)\xi \ne 0 \in L^p(\mu)$. Hence, if for some $x \in X$ we have $d\iota_p(x)\xi = 0 \in L^p(\mu)$, then $\langle \xi, x - z \rangle = 0$ for all $z \ne x$. This means that $\xi \perp \mathcal{L}(X)$, where $\mathcal{L}(X) = \text{span}\,\{x - y \,|\, x, y \in X\}$, and so $d\iota_p(x)$ is injective on $\mathcal{L}(X)$. Furthermore, we see that

$$\|d\iota_p(x)\xi\|_p^p = \int_X \left| \frac{\langle \xi, x - z \rangle}{\|x - z\|} \right|^p d\mu(z) \le \|\xi\|^p \mu(X)$$

implying that $d\iota_p(x)$ is bounded on $\mathbb{H}$.

We next show that for any $x \in X$, the map $d\iota_p(x)^{-1}$ is bounded on $d\iota_p(x)C_x^0 X$ by showing that there exists a constant $K > 0$ such that for any $x \in X$ and any $\xi \in C_x^0 X$ we have $\|d\iota(x)\xi\|_p \ge K\|\xi\|$. If not, there are sequences $\{x_n\}_{n\ge 1} \subset X$ and $\{\xi_n\}_{n\ge 1} \subset C_x^0 X$ such that $\|d\iota_p(x_n)\xi_n\|_p \le \frac{1}{n}\|\xi_n\| \le \frac{1}{n}$ for all $n \ge 1$. Since X is compact, we may as well assume that the sequence $\{x_n\}_{n\ge 1}$ converges in norm to some $x \in X$. Also, since X is spherically compact, it follows that we may assume the sequence $\{\xi_n\}_{n\ge 1}$ converges in norm to some $\xi \in C_x^0 X$. (Of course, we have, by definition, $\|\xi\| = 1$.) Consequently, for each $z \in X$ with $z \ne x$ we have that

$$\lim_{n\to\infty} \frac{\langle \xi_n, x_n - z \rangle}{\|x_n - z\|} = \frac{\langle \xi, x - z \rangle}{\|x - z\|}$$

implying that

$$\|d\iota_p(x)\xi\|_p^p \le \lim_{n\to\infty} \int_X \left| \frac{\langle \xi_n, x_n - z \rangle}{\|x_n - z\|} \right|^p d\mu(z) = 0\,.$$

Hence, the function $\langle \xi, x - z \rangle / \|x - z\| = 0$ for all $z \in X$ with $z \ne x$ since it is a continuous function of z there. But this implies that $\|\xi\| = 0$ which is a contradiction.

That the map $\iota_p\colon X \longrightarrow L^p(\mu)$ is Lipschitz is clear. We finish the proof by showing that ι_p is Lipschitz from below. If not, then there exist sequences $\{y_n\}_{n\geq 1}$ and $\{z_n\}_{n\geq 1}$ in X, with $y_n \neq z_n$, converging to some point $x \in X$ (recall that X is compact) and such that for each $n \geq 1$ we have $\|\iota_p(y_n) - \iota_p(z_n)\|_p \leq \frac{1}{n}\|y_n - z_n\|$. Since X is spherically compact, there is a subsequence of $\{(y_n - z_n)/\|y_n - z_n\|\}_{n\geq 1}$ that converges in norm to some $\xi \in C^0_x X$ with $\|\xi\| = 1$. But

$$\frac{\iota_p(y_n) - \iota_p(z_n)}{\|y_n - z_n\|} = d\iota_p(z_n)\frac{y_n - z_n}{\|y_n - z_n\|} + O\left(\|y_n - z_n\|\right)$$

and, since the map ι_p is C^1, we see that

$$\lim_{n\to\infty} d\iota_p(z_n)\frac{y_n - z_n}{\|y_n - z_n\|} = d\iota_p(x)\xi.$$

However, we also have

$$\lim_{n\to\infty}\frac{\|\iota_p(y_n) - \iota_p(z_n)\|_p}{\|y_n - z_n\|} \leq \lim_{n\to\infty}\frac{1}{n} = 0$$

by assumption. Therefore, we must have $\|d\iota_p(x)\xi\|_p = 0$ and this contradiction completes the proof. $\square$

Proof of Corollary 2.2. By Theorem 2.1 we have that the map $\iota_p\colon X \longrightarrow \iota_p(X)$ is a homeomorphism (in fact, bi-Lipschitz). The only remaining step in proving that $\iota_p\colon X \longrightarrow L^p(\mu)$ is an immersion is to show that the linear map

$$d\iota_p(x)^{-1}\colon d\iota_p(x)\mathcal{L}(X) \longrightarrow \mathcal{L}(X)$$

is bounded. To this end, we show that there is a constant $K > 0$ such that $\|d\iota_p(x)\eta\|_p \geq K\|\eta\|$ for all $x \in X$ and all $\eta \in \mathcal{L}(X)$. If not, there exist sequences $\{x_n\}_{n\geq 1} \subset X$ and $\{\eta_n\}_{n\geq 1} \subset \mathcal{L}(X)$ such that $\|\eta_n\| = 1$ and $\|d\iota_p(x_n)\eta_n\|_p < 1/n$ for all $n \geq 1$. Hence,

$$\int_X \left|\frac{\langle \eta_n, x_n - z\rangle}{\|x_n - z\|}\right|^p d\mu(z) = \|d\iota_p(x_n)\eta_n\|_p^p < \frac{1}{n^p}$$

implying that

$$\liminf_{n\to\infty}\int_X \left|\frac{\langle \eta_n, x_n - z\rangle}{\|x_n - z\|}\right|^p d\mu(z) = 0.$$

It then follows from Fatou's lemma that

$$\int_X \liminf_{n\to\infty}\left|\frac{\langle \eta_n, x_n - z\rangle}{\|x_n - z\|}\right|^p d\mu(z) = 0.$$

But, since $\mathcal{L}(X)$ is finite dimensional, we may as well assume that the sequence $\{\eta_n\}_{n\geq 1}$ converges to some $\eta \in \mathcal{L}(X)$ with $\|\eta\| = 1$. Also, since X is compact, we may assume that the sequence $\{x_n\}_{n\geq 1}$ converges to some $x \in X$. Consequently, we see that

$$\int_X \left|\frac{\langle \eta, x - z\rangle}{\|x - z\|}\right|^p d\mu(z) = \int_X \lim_{n\to\infty}\left|\frac{\langle \eta_n, x_n - z\rangle}{\|x_n - z\|}\right|^p d\mu(z) = 0$$

and so $\langle \eta, x - z\rangle = 0$ for all $z \in X$ with $z \neq x$ which implies that $\eta \perp (x - z)$ for all $z \in X$ with $z \neq x$. Because for any $z_1, z_2 \in X$ we may write $z_1 - z_2 = (x - z_2) - (x - z_1)$, we have that $\eta \perp \mathcal{L}(X)$. But $\eta \in \mathcal{L}(X)$ and so $\eta = 0$ which is a contradiction. $\square$

Using the same techniques as in the proof of Theorem 2.1, we obtain a parallel result in which the ambient space is a Riemannian manifold M instead of a Hilbert space $\mathbb{H}$.

Theorem 2.5. *Let M be a C^2 Riemannian manifold with associated metric d and without conjugate points. Let X be a compact subset of M and let μ be a finite regular Borel measure on M with closed support X. If $1 \le p < \inf_{x \in X} \underline{D}(x) < \infty$, then the map $\iota_p \colon M \longrightarrow L^p(\mu)$ given by $\iota_p(x) = d(x, \ )$ is C^1 and restricts to a C^1 embedding of X.*

3. $\mathcal{G}$-Equivariant Standard C^1 Category

In this section we give a proof of the Main Theorem. We single out the *Standard C^1 Category* as the one with objects subsets of Euclidean space. Although it is known that the compact C^1-homogeneous objects in this category are C^1-submanifolds [16], it is still interesting that the compact group objects of this category are actually Lie groups (Corollary 3.2).

If $\mathcal{G}$ is a compact Lie group and X is a compact member of the Standard C^1 Category, it makes sense to speak of a C^1 action $\mathcal{G} \times X \longrightarrow X$. In this case, the classical smooth equivariant embedding theorems of Mostow [10], Palais [15], Wasserman [17], and Bredon [1] may be readily extended to X. Once such an extension has been made, other questions become more interesting; for instance the following one:

Question. *Let $\mathcal{G}$ be a compact topological group, and suppose that $\mathcal{G} \times X \longrightarrow X$ is a continuous action via diffeomorphisms on a standard object X. Is $\mathcal{G}$ then a Lie group?*

We provide a partial answer to this question which depends on the (fairly weak) Lipschitz relation among the diffeomorphisms induced by $\mathcal{G}$ on X. For any $g \in \mathcal{G}$, we are assuming the map $g \colon x \mapsto g\,x$ is a C^1 diffeomorphism of X. Consequently, it is bi-Lipschitz (recall that X is compact). Thus we have lower and upper Lipschitz constants $\lambda(g)$ and $\Lambda(g)$ associated with that map. We now express our hypotheses in terms of $\Lambda(g)$.

Lemma 3.1. *Let $\mathcal{G}$ be a compact topological group and let $\mathcal{G} \times X \longrightarrow X$ be an effective action of $\mathcal{G}$ on a compact standard object X. Suppose that:*

(1) *For each $g \in \mathcal{G}$, the map $g \colon x \mapsto g\,x$ is a C^1 diffeomorphism of X.*
(2) *There is a constant $K > 0$ such that $\Lambda(g) < K$ for every $g \in \mathcal{G}$.*

Then we have the following:

(1) *There is an equivariant C^1 embedding of X into a finite dimensional unitary representation space of $\mathcal{G}$.*
(2) *$\mathcal{G}$ is a Lie group.*
(3) *The action $\mathcal{G} \times X \longrightarrow X$ is C^1.*

Proof. Let $L^2(\mathcal{G} \colon \mathbb{R}^n)$ denote the Hilbert space of square integrable functions on $\mathcal{G}$ with values in $\mathbb{R}^n$. We define a map $j \colon X \longrightarrow L^2(\mathcal{G} \colon \mathbb{R}^n)$ by setting $j(x)(g) = g^{-1}x$. If we let $\mathcal{G}$ act on $L^2(\mathcal{G} \colon \mathbb{R}^n)$ to define the regular representation, $g\,f(h) = f(g^{-1}\,h)$, then j is equivariant. Using the definition of tangent space for X and the associated implicit function theorem in [8] or [4], we see that j is actually a C^1 embedding. Hence, j is a $\mathcal{G}$-equivariant C^1 embedding into $L^2(\mathcal{G} \colon \mathbb{R}^n)$. (Wasserman [17] proves this fact–in much the same way–for the case of X a smooth manifold.)

Now we apply the classical Whitney trick, using the machinery of [8] or [4], to our embedding j. We know that $L^2(\mathcal{G} \colon \mathbb{R}^n)$ decomposes as a direct sum of finite dimensional unitary representations, so we may find an increasing sequence $V_1 \subset V_2 \subset \ldots$ of finite dimensional representation subspaces of $L^2(\mathcal{G}\colon \mathbb{R}^n)$, with dense union. Let $\pi_n\colon L^2(\mathcal{G}\colon \mathbb{R}^n) \longrightarrow V_n$ be the orthogonal (and equivariant) projection. Let $x \in X$ and let $K_n(x)$ be the kernel of $d\,(\pi_n \circ j)\,(x)$. Clearly we have a decreasing sequence $\cdots \supset K_n(x) \supset K_{n+1}(x) \supset \ldots$ with $\bigcap_{n \geq 1} K_n(x) = \{0\}$. Consequently, there is a finite $n(x)$ such that for $n(x) \leq n$ the map $\pi_n \circ j$ is an immersion in a neighborhood of x. Because we have assumed, for simplicity, that X is compact, we see that for $n_0 \leq n$, for some finite n_0, the composition $\pi_n \circ j$ is an immersion. It follows as usual that there is an open neighborhood U of the diagonal $\Delta(X)$ of X in $X \times X$ so that

$$\left[(\pi_n \circ j) \times (\pi_n \circ j)\right]^{-1}\left[\Delta\left(L^2(\mathcal{G}\colon \mathbb{R}^n)\right) \cap U\right] = \Delta(X) \ .$$

Now let

$$E_n = \{(x \times y) \in X \times X \,|\, x \neq y \text{ and } \pi_n \circ j(x) = \pi_n \circ j(y)\} \ .$$

We see that for $n_o \leq n$, E_n is a closed subset of $(X \times X) \setminus U$, and so it is compact. Because the sets E_n, $n \geq 1$, form a decreasing sequence with empty intersection, there exists $n_1 \geq n_0$ such that $E_n = \emptyset$ for $n \geq n_1$. This proves assertion (1).

Since $\mathcal{G}$ acts effectively on X, the representation of $\mathcal{G}$ in V_n, for $n_1 \leq n$, must be faithful. Then $\mathcal{G}$ is a closed subgroup of the rotation group of V_n and hence it is a Lie group. Finally, because the action $\mathcal{G} \times V_n \longrightarrow V_n$ is linear, the action $\mathcal{G} \times X \longrightarrow X$ must be C^1, and the proof is complete. $\square$

The following corollary is closely related to the main result of [16].

Corollary 3.2. *If $\mathcal{G}$ is a compact group object in the Standard C^1 Category, then $\mathcal{G}$ is a Lie group.*

Proof. We have C^1 maps $\mathcal{G} \times \mathcal{G} \overset{\cdot}{\longrightarrow} \mathcal{G}$ and $\mathcal{G} \overset{\mathrm{inv}}{\longrightarrow} \mathcal{G}$ which turn $\mathcal{G}$ into a topological group. Then the left translations are C^1 and, because multiplication is Lipschitz, they have a common Lipschitz bound. $\square$

The above proof also yields the following corollary.

Corollary 3.3. *Let $\mathcal{G}$ be a compact group and let X be a compact standard object. Suppose that $\mathcal{G} \times X \longrightarrow X$ is a Lipschitz action such that each translation map $g\colon X \longrightarrow X$ is a C^1 map. Then $\mathcal{G}$ is a Lie group and the action $\mathcal{G} \times X \longrightarrow X$ is C^1.*

We note that, as a consequence of Lemma 3.1, we may regard the $\mathcal{G}$-space $\mathcal{G} \times X \longrightarrow X$ as an invariant subset of a $\mathcal{G}$-manifold $\mathcal{G} \times \mathbb{R}^n \longrightarrow \mathbb{R}^n$, so that the results of [17] now transfer to $\mathcal{G} \times X \longrightarrow X$ almost verbatim.

We next apply the device of local canonical maps, provided by Theorem 2.5, to show that certain isometries in a compact Riemannian manifold are C^1 maps. More specifically, a Riemannian metric on compact smooth manifold M determines a *smooth* metric d on M, and one may ask whether an isometry $\varphi\colon M \longrightarrow M$ *with respect to the smooth metric d* is a C^1 map – of course an isometry with respect to the Riemannian metric is at least C^1 by definition. More generally now, we may consider a closed subset X of M and an isometric injection $\varphi\colon X \longrightarrow M$, and ask whether φ is C^1. In response, we have the following easily proved result.

Lemma 3.4. *Let M be a compact Riemannian manifold with corresponding smooth metric d. Let X be a closed subset of M and let μ be a geometric measure on X of dimension $D > 2$. If $\varphi\colon X \longrightarrow M$ is an isometry with respect to d, then φ is a C^1 map.*

Proof. Let $x \in X$ and let $y = \varphi(x)$. There exist open neighborhoods (in M) U and V of x and y respectively, such that :

(1) No two points in U are conjugate.
(2) No two points in V are conjugate.
(3) $\varphi(X \cap U) \subset V$.

By Theorem 2.5, the two canonical maps $\iota_2\colon X \cap U \longrightarrow L^2(\mu \,|\, (X \cap U))$ and $\iota_2'\colon \varphi(X \cap U) \longrightarrow L^2(\varphi_*[\mu \,|\, (X \cap U)])$ are C^1 embeddings because both $X \cap U$ and $\varphi(X \cap U)$ are conjugacy-free. Furthermore, the map $\varphi^*\colon L^2(\varphi_*[\mu \,|\, (X \cap U)]) \longrightarrow L^2(\mu \,|\, (X \cap U))$ is a (unitary) bounded linear transformation and the following diagram commutes.

$$
\begin{array}{ccc}
X \cap U & \xrightarrow{\;\iota_2\;} & L^2(\mu \,|\, (X \cap U)) \\
\varphi \downarrow & & \uparrow \varphi^* \\
\varphi(X \cap U) & \xrightarrow[\;\iota_2'\;]{} & L^2(\varphi_*[\mu \,|\, (X \cap U)])
\end{array}
$$

Then the composition $(\varphi^* \circ \iota_2') \circ \varphi = \iota_2$ is C^1. Also, the map $\varphi^* \circ \iota_2'\colon \varphi(X \cap U) \longrightarrow \varphi^* \circ \iota_2'\, (\varphi(X \cap U))$ is a C^1 diffeomorphism because φ^* is a C^1 diffeomorphism and ι_2' is a C^1 embedding. Consequently, the map

$$(\varphi^* \circ \iota_2')^{-1} : \varphi^* \circ \iota_2'\, (\varphi(X \cap U)) \longrightarrow \varphi(X \cap U)$$

is C^1 and hence, the composition $(\varphi^* \circ \iota_2')^{-1} \circ (\varphi^* \circ \iota_2' \circ \varphi) = \varphi$ is a C^1 map. $\square$

Corollary 3.5. *Let M be a compact Riemannian manifold with corresponding smooth metric d. Let X be a closed subset of M and let μ be a geometric measure supported on X. If $\varphi\colon X \longrightarrow M$ is an isometric embedding, then φ is C^1*

Proof. Lemma 3.4 implies that $X \times S^3 \xrightarrow{\varphi \times \mathrm{id}} M \times S^3$ is C^1 and the corollary follows immediately. $\square$

Finally, the Main Theorem follows from Lemma 3.1 and Corollary 3.5; that the group is compact, is a consequence of equicontinuity.

4. More on Canonical Embeddings

While the results of this section are not used in the proof of the Main Theorem, they are useful for filtering out pathology and therefore are of independent interest.

It is shown in [11] that a compact smoothly finite dimensional (see [8] for definition) subset of a Hilbert space need not admit a C^1 embedding into a Euclidean space. This example generalizes to a systematic family of compact smoothly finite dimensional subsets of L^p spaces, none admitting a C^1 embedding into a Euclidean space. The theorem that does the work is the following; recall that a metric d is an ultrametric on a space X if $d(x_1, x_2) \leq \max\{d(x_1, x_3), d(x_2, x_3)\}$ for every $x_1, x_2, x_3 \in X$.

Theorem 4.1. *Suppose that* (X, d, μ) *represents a space, a metric and a measure with d ultrametric and* μ *geometric of dimension D. If* $D < p$ *and* $1 \leq p$, *then for* $s = 1 - D/p$ *the map* $\iota_p^s \colon X \longrightarrow L^p(\mu)$ *defined by* $\iota_p^s(x) = d(x, \ )^s$ *is a locally bi-Lipschitz embedding.*

The image $\iota_p^s(X)$ is *never* spherically compact and *always* zero dimensional; see [11] for details. Spherical compactness is an invariant of the C^1 Category, so that no subset $\iota_p^s(X) \subset L^p(\mu)$ may be C^1 embedded in a finite dimensional Euclidean space. However, the peculiarities of ι_p^s and $\iota_p^s(X)$ go far beyond this fact: Generalizing [11], if X is of finite *metric dimension* (see [7]), we may reverse any bi-Lipschitz embedding $e \colon X \longrightarrow \mathbb{R}^n$ of an ultrametric space X as above into Euclidean space to obtain $f = \iota^s \circ e^{-1} \colon e(X) \longrightarrow L^p(\mu)$. This map is always bi-Lipschitz and *nowhere* differentiable (with respect to the norm topology on $L^p(\mu)$). For n larger than the metric dimension of X, the embedding $e \colon X \longrightarrow \mathbb{R}^n$ always exists by [7, Theorem 3.8]. Even more peculiar, any smoothly embedded compact finite dimensional submanifold of $L^p(\mu)$ can intersect $\iota_p^s(X)$ in at most a finite set.

For the general case, the following proposition is not hard to prove.

Proposition 4.2. *If* (X, d, μ) *represents a space, a metric and a measure and if* $a\, r^D \leq \mu\left(B(x, r)\right)$ *for all* $x \in X$ *and* $0 \leq r \leq r_0$, *then for* $1 \leq p$, *the map* $\iota_p \colon X \longrightarrow L^p(\mu)$ *is Lipschitz and* $(\iota | X)^{-1} \colon \iota(X) \longrightarrow X$ *is locally Hölder of order* $p/(p + D)$.

We finish this section with a proof of Theorem 4.1.

Proof of Theorem 4.1. Let $0 < r < 1/2$ be fixed and let $x, y \in X$ with $d(x, y) = r^n$ for some $n \geq 1$. Then

$$\|\iota_p(x) - \iota_p(y)\|_p^p = \int_{B(x, r^n)} |\, d(x, z) - d(y, z)|^p \, d\mu(z)$$

$$+ \int_{X \setminus B(x, r^n)} |\, d(x, z) - d(y, z)|^p \, d\mu(z) \, ,$$

where $B(x, r^n)$ denotes the closed ball of radius r^n centered at x. Since (X, d) is an ultrametric space, we may assume without loss of generality [7, Theorem 2.2 and Remarks 2.3], that the range of d is in the set $\{0\} \cup \{r^n\}_{n \geq 1}$. Observe that if $z \in X \setminus B(x, r^n)$, then $z \in X \setminus B(y, r^n)$ and so $d(x, z) = d(y, z)$ implying that

$$\int_{X \setminus B(x, r^n)} |\, d(x, z) - d(y, z)|^p \, d\mu(z) = 0 \, .$$

Furthermore, if $z \in B(x, r^{n+1})$, then $d(y, z) = r^n$. Similarly, if $z \in B(y, r^{n+1})$, then

$d(x, z) = r^n$. Hence, we have

$$\|\iota_p(x) - \iota_p(y)\|_p^p = \int_{B(x,r^n)} |d(x,z) - d(y,z)|^p \, d\mu(z)$$

$$= \int_{B(x,r^{n+1})} |d(x,z) - r^n|^p \, d\mu(z)$$

$$+ \int_{B(y,r^{n+1})} |d(y,z) - r^n|^p \, d\mu(z)$$

$$= \sum_{j=1}^{\infty} \int_{B(x,r^{n+j})\setminus B(x,r^{n+j+1})} [r^n - d(x,z)]^p \, d\mu(z)$$

$$+ \sum_{j=1}^{\infty} \int_{B(y,r^{n+j})\setminus B(y,r^{n+j+1})} [r^n - d(y,z)]^p \, d\mu(z) \ .$$

For each $j \geq 1$ we see that

$$\int_{B(x,r^{n+j})\setminus B(x,r^{n+j+1})} [r^n - d(x,z)]^p \, d\mu(z)$$
$$= (r^n - r^{n+j})^p \mu \left[B(x,r^{n+j}) \setminus B(x,r^{n+j+1}) \right]$$

which implies that

$$\|\iota_p(x) - \iota_p(y)\|_p^p = r^{np} \sum_{j=1}^{\infty} \left(1 - r^j\right)^p \mu \left[B(x,r^{n+j}) \setminus B(x,r^{n+j+1}) \right]$$

$$+ r^{np} \sum_{j=1}^{\infty} \left(1 - r^j\right)^p \mu \left[B(y,r^{n+j}) \setminus B(y,r^{n+j+1}) \right] \ .$$

Since $0 < r < 1/2$, it is immediate that

$$2^{-p} \mu \left[B(x,r^{n+1}) \right] \leq \sum_{j=1}^{\infty} \left(1 - r^j\right)^p \mu \left[B(x,r^{n+j}) \setminus B(x,r^{n+j+1}) \right] \leq \mu \left[B(x,r^{n+1}) \right]$$

and thus

$$2^{-p} r^{np} \left\{ \mu \left[B(x,r^{n+1}) \right] + \mu \left[B(y,r^{n+1}) \right] \right\} \leq \|\iota_p(x) - \iota_p(y)\|_p^p$$
$$\leq r^{np} \left\{ \mu \left[B(x,r^{n+1}) \right] + \mu \left[B(y,r^{n+1}) \right] \right\}.$$

Because μ is a geometric measure, for any $z \in X$ and for n sufficiently large, we have the inequalities $a\, r^{D(n+1)} \leq \mu\left(B(z,r^{n+1})\right) \leq A\, r^{D(n+1)}$ for some constants $0 < a < A$ and so

$$a\, r^D r^{n(D+p)} \leq \|\iota_p(x) - \iota_p(y)\|_p^p \leq 2^p A\, r^D r^{n(D+p)} \ .$$

But $d(x,y) = r^n$ and hence for x and y sufficiently close (i.e., n sufficiently large) we have

$$K_1\, d(x,y)^{1+D/p} \leq \|\iota_p(x) - \iota_p(y)\|_p \leq K_2\, d(x,y)^{1+D/p},$$

where $K_1 = (a \, r^D)^{1/p}$ and $K_2 = 2(A \, r^D)^{1/p}$. Therefore, the map $\iota_p \colon X \longrightarrow L^p(\mu)$ is locally bi-Hölder of order $1 + D/p$.

Now since d is an ultrametric, then d^t is also an ultrametric for any $t > 0$. Hence, there exists $s > 0$ (to be determined) such that the map ι_p is locally bi-Lipschitz with respect to the ultrametric $\delta = d^s$. Let $B(x, r)$, respectively $B_\delta(x, r)$, denote the closed ball of radius r centered at $x \in X$ in the metric d, respectively in the metric $\delta = d^s$. Then $B(x, r) = B_\delta(x, r^s)$ and consequently we have $a \, (r^s)^{D/s} \le \mu \left(B_s(x, r^s)\right) \le A \, (r^s)^{D/s}$. Therefore we have

$$K_1 \, d(x, y)^{s(1 + D/sp)} = K_1 \, \delta(x, y)^{1 + D/sp} \le \|\iota_p^s(x) - \iota_p^s(y)\|_p$$
$$\le K_2 \, \delta(x, y)^{1 + D/sp} = K_2 \, d(x, y)^{s(1 + D/sp)}.$$

By setting $s = 1 - D/p$ we see that

$$K_1 \, d(x, y) \le \|\iota_2^s(x) - \iota_p^s(y)\|_p \le K_2 \, d(x, y)$$

for any $x, y \in X$ with $d(x, y)$ sufficiently small. This completes the proof. $\square$

5. Remarks and Examples

i) Considering the Main Theorem, we point out that the *arclength metric* on fairly nice subsets of Euclidean space produces isometries which are not C^1 and fail to form a finite dimensional Lie group:

Example 5.1. Let X be the closure of the set obtained by rotating the graph of $y = x^2 \sin^2(1/x)$ with $0 < x \le 1/\pi$ about the x–axis. With $\mu =$ surface measure, the goup of isometries is clearly $O(2) \times O(2) \times O(2) \times \cdots$ most of which are not C^1; The group of C^1 isometries is just $O(2)$. And then the Main Theorem implies that the canonical map $\iota_2 \colon X \longrightarrow L^2(\mu)$ cannot be C^1.

ii) The following counterexample shows that the hypothesis that X is a standard object (spherical compactness) cannot be deleted from Lemma 3.1. However, we do not know whether this counterexample is a C^1 group object or not, and so we cannot address Corollary 3.2.

Example 5.2. Let $\mathcal{G} = X = (\mathbb{Z}_2)^\infty$ regarded as a topological group with Haar measure μ. We may impose an invariant ultrametric d on X so that the measure μ becomes a finite dimensional geometric measure, of say dimension $D < 2$. Let $s = 1 - D/2$ and let $\iota_2^s \colon X \longrightarrow L^2(\mu)$ be the map $\iota_2^s(x) = d(x, \ )^s$. Then, by Theorem 4.1 and [11] , the map ι_2^s is a bi-Lipschitz embedding with $\iota_2^s(X)$ smoothly zero dimensional (see [8] for definition), but *not* spherically compact. For the obvious action $\mathcal{G} \times \iota_2^s(X) \longrightarrow \iota_2^s(X)$ we have that each homeomorphism $g \colon \iota_2^s(X) \longrightarrow \iota_2^s(X)$ extends to a unitary isomorphism of $L^2(\mu)$. Thus $\mathcal{G}$ acts on X via C^1 diffeomorphisms, and they have the common bound $K = 1$ for their Lipschitz constants. However, $\mathcal{G}$ is clearly not a Lie group, so that the hypothesis of spherical compactness cannot be dropped, and Lemma 3.1 is sharp.

iii) In [8, Theorem 5.3] necessary and sufficient conditions are given for a compact subset of a Hilbert space to admit a C^1 embedding into a Euclidean space.

REFERENCES

1. G. E. Bredon, *Introduction to Compact Transformation Groups*, Academic Press, New York, 1972.
2. J. Dugundji, *Topology*, Allyn and Bacon, Boston, 1966.
3. K. J. Falconer, *The Geometry of Fractal Sets*, Cambridge University Press, Cambridge, 1985.
4. G. Glaeser, *Étude de quelque algèbres tayloriennes*, J. Analyse Math. **6** (1958), 1–124, erratum, insert to **6** (1958), no. 2.
5. M. Gromov, *Filling Riemannian manifolds*, J. Differential Geom. **18** (1983), 1–147.
6. A. Jonsson and H. Wallin, *Function Spaces on Subsets of* $\mathbb{R}^n$, Math. Rep. (Chur, Switzerland) (J. Peetre, ed.), Vol. 2, pt. 1, Harwood Academic, New York, 1984.
7. J. Luukkainen and H. Movahedi-Lankarani, *Minimal bi-Lipschitz embedding dimension of ultrametric spaces*, Fund. Math. **144** (1994), 181–193.
8. R. Mansfield, H. Movahedi-Lankarani, and R. Wells, *Smooth finite dimensional embeddings*, Manuscript, 1996.
9. J. Milnor, *Lectures on Differential Topology*, Princeton University, 1958.
10. G. D. Mostow, *Equivariant embeddings in Euclidean space*, Ann. of Math. **65** (1957), 432–446.
11. H. Movahedi-Lankarani, *On the theorem of Rademacher*, Real Analysis Exchange **17** (1992), 802–808.
12. H. Movahedi-Lankarani and R. Wells, *Ultrametrics and geometric measures*, Proc. Amer. Math. Soc. **123** (1995), 2579–2584.
13. H. Movahedi-Lankarani and R. Wells, *On geometric quotients*, Manuscript, 1996.
14. K. Nomizu, *Lie Groups and Differential Geometry*, Publications of Math. Soc. of Japan, 1956.
15. R. S. Palais, *Imbedding of compact differentiable transformation groups in orthogonal representations*, J. Math. Mech. **6** (1975), 673–678.
16. D. Repovš, A. B. Skopenkov, and E. V. Ščepin, C^1-*homogeneous compacta in* $\mathbb{R}^n$ *are* C^1-*submanifolds of* $\mathbb{R}^n$, Proc. Amer. Math. Soc. **124** (1996), 1219–1226.
17. A. G. Wasserman, *Equivariant differential topology*, Topology **8** (1969), 127–150.

DEPARTMENT OF MATHEMATICS, PENN STATE ALTOONA, ALTOONA, PA 16601-3760
E-mail address: hml@math.psu.edu

DEPARTMENT OF MATHEMATICS, PENN STATE UNIVERSITY, UNIVERSITY PARK, PA 16802
E-mail address: wells@math.psu.edu